D0098463

The Cosmic Web

b

Mysterious Architecture of the Universe

J. Richard Gott

PRINCETON UNIVERSITY PRESS

Princeton and Oxford

Published by Princeton University Press, 41 William Street, Princeton, New Jersey 08540
In the United Kingdom: Princeton University Press, 6 Oxford Street, Woodstock, Oxfordshire OX20 1TW
press.princeton.edu

Jacket image courtesy of Illustris Collaboration

ISBN 978-0-691-15726-9

Library of Congress Cataloging-in-Publication Data

Gott, J. Richard.
The cosmic web : mysterious architecture of the universe / J. Richard Gott.
Princeton : Princeton University Press, [2016] | Includes bibliographical references and index.
LCCN 2015026434 | ISBN 9780691157269 (alk. paper)
LCSH: Astronomers—Biography. | Gott, J. Richard. | Astronomy—History. | Cosmology.
LCC QB35 .G68 2016 | DDC 523.1—dc23 LC record available at http://lccn.loc.gov/2015026434
British Library Cataloging-in-Publication Data is available

This book has been composed in Minion Pro and ITC Goudy Sans

Printed on acid-free paper ∞

Printed in the United States of America

10 9 8 7 6 5 4 3 2 1

To Mrs. Ruth Pardon, my high-school math teacher; Dr. Bruce Wavell, head of the Rollins College summer math program; Mrs. Dorothy Schriver, Science Talent Search Program Manager; Drs. James E. Gunn and Martin Rees—who all set me on my way; and to my colleagues in the topology group, who accompanied me on our journey through the cosmic web. Finally, to my new granddaughter Allison—welcome to the universe.

Contents

Acknowledgments ix

Preface xi

Chapter 1. Hubble Discovers the Universe 1

Chapter 2. Zwicky, Clusters of Galaxies, and the Discovery of Dark Matter 28

Chapter 3. How Clusters Form and Grow—Meatballs in Space 41

Chapter 4. The Great Void in Boötes—A Swiss Cheese Universe 64

Chapter 5. Inflation 79

Chapter 6. A Cosmic Sponge 103

Chapter 7. A Slice of the Universe—the Great Wall of Geller and Huchra 135

Chapter 8. Park's Simulation of the Universe 144

Chapter 9. Measuring the Cosmic Web—the Sloan Great Wall 155

Chapter 10. Spots in the Cosmic Microwave Background 180

Chapter 11. Dark Energy and the Fate of the Universe 193

Notes 227

References 235

Index 245

Acknowledgments

First and foremost I thank my wife, Lucy, for her love and support and for her professional expertise in editing the manuscript for clarity. I thank my daughter Elizabeth and son-in-law Michael for their love and support. I thank my colleague Michael Vogeley, who kindly read the entire manuscript and offered his excellent comments and additions, Bob Vanderbei and Li-Xin Li for assistance with diagrams, and Zachary Slepian, Matias Zaldarriaga, Nima Arkani-Hamed, and Andrew Hamilton for their helpful comments. I thank my agent Jeff Kleinman, who is always a pleasure to work with, and my wonderful editor at Princeton University Press, Ingrid Gnerlich, and her assistant, Eric Henney. I thank my production editor, Brigitte Pelner, my copyeditor, Linda Thompson, and my illustration manager, Dimitri Karetnikov, for their expert help.

It is a pleasure to acknowledge the many colleagues with whom I have worked on large-scale structure in the universe: Jim Gunn, Martin Rees, Ed Turner, Sverre Aarseth, and Suketu Bhavsar early on; Adrian Melott and Mark Dickinson, with whom I developed the spongelike topology idea; Andrew Hamilton, who derived a critical formula; David Weinberg, who implemented its use in many collaborations; Trinh X. Thuan and Michael Vogeley, with whom I worked on observational samples; Changbom Park, whose computer simulations proved cold dark matter could produce Great Walls; Barbara Ryden who worked on topology in rodlike samples, Wes Colley and Changbom Park, with whom I worked on the cosmic microwave background; Mario Jurić, who helped measure the Sloan Great Wall; Lorne Hofstetter who helped make a picture of it; Juhan Kim, who, together with Changbom Park created the Horizon Run simulations; and Clay Hambrick, Yun-Young Choi, Robert

Speare, and Prachi Parihar, who took part in our project to apply our topology technique to compare the large computer simulations with observations from the Sloan Survey. I thank Zack Slepian for his collaboration on a formula to characterize dark energy. Many of these people have followed this project for decades and have become lifelong friends. I treasure all these associations.

Preface

Galileo once said: "Philosophy [nature] is written in that great book which ever is before our eyes—I mean the universe. . . . The book is written in mathematical language, and the symbols are triangles, circles and other geometrical figures." So it proved to be with the arrangement of galaxies in the universe. To understand it would require geometrical language.

When I was 18 years old, I discovered a group of intricate, spongelike structures made of triangles, squares, pentagons, or hexagons—some of which neatly divided space into two equal and completely interlocking regions. These were regular spongelike polyhedrons—figures composed of regular polygons whose arrangement around each vertex was identical. Being a teenager, when confronted with the ancient Greek wisdom that there were five, and only five, regular polyhedrons (the tetrahedron, cube, octahedron, dodecahedron, and icosahedron)—and that this had been proven long ago—I said, "Well, maybe not." I made this my high-school science project and took it to my local science fair in Louisville, Kentucky. Surprisingly, this would later play a role in my own path to understanding the arrangement of galaxies in the universe.

Johannes Kepler was my inspiration. He had also questioned the ancient wisdom of the five regular polyhedrons. Kepler thought that the three regular polygonal tilings of the plane should be counted as polyhedrons also: the checkerboard, the hexagonal chicken-wire pattern, and triangles, six around a point, filling the Euclidean plane. Both the checkerboard and the cube were equally regular arrangements of polygons (even though one turned out flat and the other, three-dimensional). Kepler thought a checkerboard, for example, could be considered a new regular polyhedron—with an infinite number of faces. But Kepler didn't stop there; he also recognized two new regular *starred* polyhedrons. One has faces that are five-pointed stars like those on the American flag. Isn't a star just as regular as a pentagon? It has five points, just like the pentagon, and is likewise made by drawing five equal-length lines connecting them. The only difference is that the lines are allowed to cross through each other! You just have to expand your mind a little to see five-pointed stars as regular. Kepler would take five-pointed stars as the faces of his new regular polyhedron. He had them cross through each other to form a three-dimensional star. Kepler understood that you could find new things by breaking the rules just a little. (See Color Plate 1.)

Kepler was also fascinated with how one might use polyhedrons in astronomy. There were six known planets in his day. If you built a set of spheres whose radii marked the distances of each from the Sun, you would have six nested spheres. He thought that you might fit the five previously known regular polyhedrons between each of these spheres to explain the geometry of the solar system. In this he was wrong. And when more planets were discovered, the idea broke down completely. But when Kepler was told planets must have circular orbits, he thought to use elliptical orbits instead, and in this he was famously right.

But would my spongelike polyhedrons—which had geometries like a marine sponge, with many holes percolating through them—remain a mathematical fantasy, or would they ever have any practical application in real-world astronomy? It turned out they had an application in understanding galaxy clustering.

Edwin Hubble discovered that our Milky Way galaxy containing 300 billion stars was not alone in space. There were countless other galaxies just as big as ours. Furthermore, this whole assembly of galaxies was expanding, as I describe in Chapter 1. But how exactly are these galaxies arranged in space? It was a puzzle that confronted astronomers. Galaxies congregated in clusters. Chapter 2 tells how Fritz Zwicky famously studied this at Caltech. His work led American cosmologists during the Cold War to adopt a meatball model in which the high-density clusters floated in a low-density sea, as described in Chapter 3. But the Russian school of cosmology favored a model where galaxies traced a giant honeycomb in space with large empty isolated voids. This was a Swiss cheese universe (Chapter 4). I found that the new theory of inflation[1] (Chapter 5) was inconsistent with either of these pictures and required a spongelike structure in which great clusters of galaxies were connected by filaments of galaxies and great voids were connected to each other by low-density tunnels (Chapter 6).

Considering the theory of inflation and remembering those polyhedrons from my youth, I wrote a paper with Adrian Melott (University of Kansas) and Mark Dickinson (Princeton University) predicting that galaxies must be arranged on a giant cosmic sponge. The efforts we made to verify this prediction became part of the larger story of how teams of observers embarked on heroic efforts to map the universe, as described in Chapters 7, 8, and 9. These studies would give us vital insight into how the universe began. Astronomers began to chart the distribution of galaxies in space. Just as cartographers of the past mapped Earth, these cosmic cartographers began mapping our universe. Starting with surveys of a thousand galaxies, major surveys have now grown to encompass well over a million galaxies. Three-dimensional maps of the galaxies' distribution have now been made, and the structure they reveal has indeed proved to be spongelike. Great clusters of galaxies are connected by *filaments*, or chains of galaxies, in a spongelike geometry, while the low-density voids are connected to each other by low-density tunnels; this entire structure is now called the *cosmic web*. Fantastic filamentary chains of galaxies connecting great clusters have been found stretching over a billion light-years in length. These are the largest structures in the universe. Measuring one of them, called the Sloan Great Wall, landed Mario Jurić and me in the *Guinness Book of Records*—and we didn't even have to collect the world's largest ball of twine! I will explain how these largest structures in the universe arose as the greatly expanded fossil remnants of microscopic random quantum fluctuations in the early universe produced by inflation in the universe's first 10^{-35} seconds. This is supported by study of the fluctuations in the cosmic microwave background radiation left over from the universe's first moments (Chapter 10).

Not only do these structures illuminate the early universe, but they can also be used to forecast our future, as described in the final chapter. Will the universe keep expanding exponentially forever, as some models suggest, or will it ultimately coast along in a slower fashion? Or, will the universe end catastrophically with a Big Rip singularity in the next

150 billion years? A careful study of the cosmic web can help answer these questions. Distinguishing among these possible alternative futures is one of the highest-priority areas of research in astronomy today.

Ranging from a humble high school science project to mapping projects involving hundreds of astronomers, this book will give you a window on how scientific research is done. It is a story of how unexpected connections can lead to new insights and how computer simulations combined with giant telescopic surveys have transformed our understanding of the universe in which we live. This is a semiautobiographical account focusing on my adventures but also emphasizing many of the people whose seminal ideas have influenced the field. I have had the good fortune to work with some of the greatest astronomers of our generation, investigating many of the aspects of this story in one way or another, from galaxy clustering, gravitational lensing, computer simulations, and mapping large-scale structures to inflation and dark energy. This book is told from my personal perspective as I meandered through the complicated web of talented people who fought for and finally won an understanding of how the universe on large scales is arranged. A cosmic web, if you will.

J. Richard Gott
Princeton, New Jersey

The Cosmic Web

Chapter 1

Hubble Discovers the Universe

It is fair to say that Edwin Hubble discovered the universe. Leeuwenhoek peered into his microscope and discovered the microscopic world; Hubble used the great 100-inch-diameter telescope on Mount Wilson in California to discover the macroscopic universe.

Before Hubble, we knew that we lived in an ensemble of stars, which we now call the Milky Way Galaxy. This is a rotating disk of 300 billion stars. The stars you see at night are all members of the Milky Way. The nearest one, Proxima Centauri, is about 4 light-years away. That means that it takes light traveling at 300,000 kilometers per second about 4 years to get from it to us. The distances between the stars are enormous—about 30 million stellar diameters. The space between the stars is very empty, better than a laboratory vacuum on Earth. Sirius, the brightest star in the sky, is about 9 light-years away.

The Milky Way is shaped like a dinner plate, 100,000 light-years across. We are located *in* this thin plate. When we look perpendicular to the plate, we see only those stars that are our next-door neighbors in the plate; most of the stars in these directions are less than a few hundred light-years away. We see about 8,000 naked-eye stars scattered over the entire sky; these are all our nearby neighbors in the plate, a tiny sphere of stars nestled within the thin width of the plate. But when we look out through the plane of the plate we see the soft glow of stars that are much farther from us but still within the plane of the plate. They trace a great circle 360° around the sky. Here we are seeing the circumference of the giant plate itself, as we look around the sky in the plane of the plate. We call this band of light the Milky Way. When Galileo looked at this band of light in his telescope in 1610, he found its faint glow was due to a myriad of faint stars—faint because they are so distant. With the naked eye we can see only their combined faint glow; we cannot resolve that glow into individual stars. It took a telescope to do that. For a long time, this constituted the known universe. Our galaxy appeared to be sitting alone in space—an island universe.

In 1918 our idea of our place in the universe started to change. Harlow Shapley discovered that the Sun was not at the center of the Milky Way but instead was about halfway out toward the edge. We were off center. Shapley felt like the new Copernicus. Just as Copernicus had moved Earth from the center of the solar system and properly placed the Sun at its center, Shapley moved the solar system from the center of the Milky Way to a place in its suburbs. Our position in the universe was looking less and less special. Shapley's monumental work did revolutionize our thinking about our place in the universe. He had a right to suppose that he had made what would be the most important discovery in astronomy in the twentieth century. *Time* would later put Shapley on its cover, on July 29, 1935. Shapley was the dean of American astronomers. But his great discovery of 1918 was soon to be eclipsed—twice—by Hubble.

Hubble studied the Andromeda Nebula, which had been thought by many, including Shapley, to be a gas cloud within the Milky Way. The word *nebula* comes from the Latin *nubes*, or "cloud," denoting the fuzzy appearance of these objects. By careful observations with the new 100-inch telescope, Hubble discovered that Andromeda was actually an entire galaxy roughly the size of the Milky Way and very far away. Furthermore there were many other similar spiral-shaped nebulae seen in the sky, and these were *all* galaxies like our Milky Way! He classified galaxies by their shapes—elliptical, spiral, and irregular—like some botanist classifying microbes. He observed in different directions and counted the number of galaxies he found. There seemed to be an equal number in different directions. On the largest scales the universe was homogeneous. There were fainter galaxies further and further away. We were just one galaxy in a vast universe of galaxies. This would have been discovery enough, but Hubble was not finished. He measured the distances to these galaxies. From spectra of these galaxies he could measure their velocities. He found that the further away a galaxy was, the faster it was moving away from us. The whole universe was expanding! This was astonishing. Isaac Newton had a static universe. Even Einstein, genius of curved spacetime, thought the universe must be static. The discovery that the universe was expanding was quite simply, astounding. It caused Einstein to revise his ideas about his field equations of general relativity—to backtrack on the changes he had made in them to produce a static cosmology. The expansion of the universe has profound implications.

If the universe were static, as Newton and Einstein had supposed, then it could be infinitely old. It would always have been here. This avoided Aristotle's problem of first cause. If the universe had a finite age, however, then *something* must have caused it. But what caused *that*? Unless one is willing to accept an infinite regression of causes, there must be a first cause—but the question remains: what caused the first cause? An expanding universe brought this question back into play. If you played the tape of history backward, you would see all the galaxies crashing together in the past. There must have been something to start all this expansion—a Big Bang—that began the universe. We now know this occurred 13.8 billion years ago. What caused this Big Bang? Astronomers following Hubble would work on that.

Hubble was the most important astronomer in the twentieth century. *Time* magazine put him on its cover on February 9, 1948. Behind him was a picture of the Palomar Observatory, whose new 200-inch-diameter telescope could extend Hubble's observations. He was the first person to observe with that telescope. Later *Time* would select Hubble as one of the 100 most influential people in the twentieth century (the only astronomer so honored). Despite the acknowledged importance of his discoveries, Hubble failed to get the American Astronomical Society's highest award, the Russell Lectureship, given each year to an outstanding American astronomer for lifetime achievement. It reminds one of the Nobel Prize committee's failure to award the Nobel Prize in Literature to Leo Tolstoy, even though they had several chances to do so before he died. The greatest people are often controversial. As with most groundbreaking discoveries, the whole story is more complicated, and interesting, than just the simple outline I have given so far. So let's look into the story in more detail.

Shapley Blazes the Trail

Harlow Shapley had measured the position of the Sun in the Milky Way by using globular clusters. He measured their distances using RR Lyrae variable stars as objects of standard luminosity—*standard candles*. RR Lyrae stars are 40 to 50 times as luminous as the Sun and so can be seen out to fairly large distances. They all have about the same intrinsic *luminosity*, the same wattage as lightbulbs, if you will. (The Sun, for example, has a luminosity of 4×10^{26} watts—equal to 4 trillion-trillion 100-watt lightbulbs.) If you saw an RR Lyrae star, you could figure out how far away it was by seeing how faint it appeared to be in the sky. It's like seeing a row of standard street lights extending down a street. They all have the same intrinsic luminosity, but the most distant ones will be fainter than the nearby ones.

Light emitted from a star spreads out in all directions, creating an ever-expanding sphere of light. Let's say you are 1,000 light-years from a star. The light that is passing you from that star is a spherical shell with a radius r of 1,000 light-years. The area of that sphere is $4\pi r^2$, or about 12 million square light-years. If you were 2,000 light-years away, the light would be diluted over an area of $4\pi r^2$ or $4\pi \times (2{,}000 \text{ light-years})^2$—about 4×12 million square light-years. The new sphere is twice as big as the one before and has an area 4 times as great. This means that your detector—let's say your 200-inch-diameter telescope—will intercept ¼ as much radiation from the star as it would if it were only 1,000 light-years away from the star. If you are twice as far away, the star appears ¼ as bright. Brightness is measured in watts per square meter falling on your detector. Brightness diminishes like one over the square of the distance, a fundamental relationship called, not surprisingly, the *inverse-square law*.

Shapley could take repeated pictures of globular clusters of stars. A globular star cluster orbiting within the Milky Way would contain over 100,000 stars orbiting about the cluster's center of mass, like bees around a hive. Stars whose brightness varied from picture to picture could be identified as variable stars. Shapley could measure these stars' brightnesses as a function of time. He could recognize RR Lyrae variables by their periods of oscillation (the length of time between peaks in brightness, characteristically less than a day) and their amplitude of oscillation (the factor by which their brightness changed from brightest to faintest). Shapley could look at a particular RR Lyrae star and know its intrinsic luminosity. This was invaluable. Knowing its intrinsic luminosity, he could measure its apparent brightness in the sky and calculate its distance. The fainter it was, the farther away it would be. By measuring the apparent brightness of the RR Lyrae variables in a globular cluster, Shapley could measure the distance to the globular cluster itself. For more distant globular clusters, he used the brightness of the brightest stars in the cluster as a distance indicator, and for the most distant globular clusters, he used the clusters' angular sizes to estimate their distances: a cluster half the angular size was twice as far away.

Shapley measured the distances to many globular clusters, which orbit the center of the Milky Way galaxy in a nearly spherical distribution along paths that take them far above and below the flat "dinner plate" where most stars lie. Looking out above and below the galactic plane allowed him to find globular clusters at great distances, free of the confusing obscuring effects of interstellar dust in the plane itself. Shapley found that the 3D distribution of globular clusters in space was off-center relative to Earth. This result was puzzling: these globular clusters were orbiting the center of the Milky Way and should be centered on it, yet Shapley found more globular clusters (and ones that were further away) on one side of the sky than on the other. The distribution of globular clusters seemed centered on a point in the direction of the constellation of Sagittarius about 25,000 light-years away. This point marked the center of the galaxy. Shapley had shown that *we* were not at the center of the Milky Way—but rather our solar system was about halfway between the center and the outer edge. This showed that the Sun was not at a special location at the center of the galaxy.

In 1920 Shapley had a famous debate with Heber Curtis about the nature of the spiral nebulae. In the period from 1771 to 1781 Charles Messier had made a catalog of nebulae. Through a small telescope they look like softly blurry patches of light and can be confused with comets. Messier was a comet hunter and wanted to make sure he didn't mistake these objects for new comets, so he took special note of them and cataloged them. These blurry objects actually include a number of different types of things. Some Messier objects (labeled by an M followed by their number in the catalog) are supernova gas ejecta (like the Crab

Nebula M1) and some, like the Dumbbell Nebula (M27), are gas shed during the process of a star collapsing to form a white dwarf. Some are globular clusters (like M13), some are loose star clusters like the Pleiades (M45), many are gas clouds (star-forming regions) in the Milky Way, like the Orion Nebula (M42), and many more are actually external galaxies, like Andromeda (M31), the Pinwheel (M101), the Whirlpool (M57), M81, M87, and so on. The spiral nebulae, such as M31, M57, M81, and M101, were the subjects of the Shapley-Curtis debate. Their spiral shapes made them look somewhat like hurricanes seen from space. They had spiral arms winding outward from the center—like a pinwheel. Sometimes they were seen face-on, where they showed off circular shapes, and sometimes they were seen nearly edge-on, looking like dinner plates seen from the side. Were these gas clouds within the Milky Way or were they external galaxies like our own seen at great distances? Shapley maintained that they were gas clouds within the Milky Way. Curtis maintained they were external galaxies just like our own.

The proposals of famous astronomers and philosophers of the past came into the mix. The ancient Greek philosopher Democritus proposed that the band of light known as the Milky Way could actually be the light of distant stars (right idea—and in about 400 BC!). This idea would be confirmed by Galileo when he turned a telescope to the heavens. In 1750 Thomas Wright speculated that the Milky Way was a thin sheet of stars (right) but thought this was really part of a large, thin spherical shell of stars orbiting a dark center (wrong). Thus from a great distance he thought our galaxy should resemble a sphere of stars, a round blurry blob. Then he proposed that many of the faint nebulae we saw were entire galaxies like our own (right!). In 1755 William Herschel (the discoverer of Uranus) designated a subclass of nebulae he called "spiral nebulae." That same year the preeminent philosopher of his day, Immanuel Kant, proposed that the spiral nebulae were actually galaxies like our own seen at great distances—he called them "island universes." Curtis had these ideas on his side.

Shapley spent most of the time defending his recent determination of the enormous size of the Milky Way; he thought this result would make the predicted distances to the spiral nebulae seem ridiculously large if they were to be objects comparable to the Milky Way in size. Some novae (stars that suddenly flare in brightness by a large factor without exploding) were observed in spiral nebulae, and these had brightnesses comparable to other novae in the Milky Way, placing them firmly within our galaxy. Curtis mentioned this point against himself. But in fact, these were *supernovae*, not novae at all but vastly more luminous stellar explosions that were actually just as far away as Curtis needed. Curtis's best argument came from noticing that the spectra of the spiral nebulae looked like the spectra of star clusters, not those of gas clouds. The debate ended inconclusively. Most people in the audience probably left with the same views they had when they entered. In science, such questions are not settled by debates or by who scores more oratorical points. They are often settled by new and decisive data—which Hubble would soon be perfectly positioned to supply.

Hubble Changes the Game

Like most people who make important contributions, Hubble was blessed with both talent and luck. Born in Marshfield, Missouri, in 1889, Hubble held the high school high-jump record for the state of Illinois. He attended the University of Illinois and later went to Oxford as a Rhodes Scholar. Rhodes scholarships rewarded athletic as well as academic prowess. When he returned from England, he spent some time in my hometown of Louisville, Kentucky, living for part of that time in a quiet, genteel area of Louisville called the Highlands, where my mother and grandmother once lived. Hubble followed his father's wishes that he study law, but after his father's death, he turned to his true interests in science. He was a high school teacher for a while before going to graduate school at the University of Chicago, where he earned his PhD in astronomy; for his thesis research, he took photographs of faint nebulae. Here he had mastered the skill that would be needed to settle the Curtis-Shapley controversy. After a brief period of service in World War I, he returned to get a staff position at Mount Wilson. He was hired by George Ellery Hale. His good fortune was compounded. Yerkes Observatory, where he had done his doctoral work, possessed the largest refracting telescope in the world with a diameter of 40 inches. This was and still remains the largest refracting telescope ever built. It had a lens at the front, which brought light to a focus at the back, where an eyepiece was placed to view the image. Galileo's first telescope was a refracting telescope whose lens had a diameter of 1.46 inches. With this he was able to resolve stars in the soft band of light called the Milky Way. The Yerkes telescope was 40 inches in diameter, or 27 times as large in diameter. A lens is like a bucket to catch light, with a light-gathering power proportional to its area. (Put a bucket out in the rain; if it has twice the diameter, its opening area will be four times as large and will collect four times as much rain.) The Yerkes telescope had 27 × 27, or 729, times the light-gathering power of Galileo's telescope. Since brightness falls off like the square of the distance, it should be able to discern stars 27 times more distant than those Galileo could see. Furthermore, long exposures using film gathered light over time and were more sensitive than the human eye. Hubble was by now an expert at taking just these kinds of pictures.

Now George Ellery Hale—telescope builder par excellence—enters. Hale, who had built the 40-inch Yerkes telescope, was just now finishing construction of the largest telescope ever—a reflecting telescope 100 inches in diameter—on Mount Wilson. A telescope with a 40-inch-diameter lens was about the largest of that type you could build. The lens had to be supported around its edge and began to sag in the middle if it was too big and heavy. But a reflecting telescope of the type invented by Isaac Newton let light come in the front and hit a big mirror in the back, where it was reflected back toward the front again. The light could then be directed via a small secondary mirror to a focus outside the tube where you would put the eyepiece. The big mirror was supported on its entire back surface and thus could be larger. The 100-inch-diameter reflecting telescope was 2.5 times the diameter of the Yerkes telescope and able to detect individual stars 2.5 times further away. If Galileo could discern individual stars that were 25,000 light-years away in the Milky Way, the 100-inch telescope should be able to detect individual stars 1.6 million light-years away. With the advantage of long-exposure photographs, Hubble could extend this distance even farther.

Hubble arrived in Los Angeles in 1919 to take up his new job soon after the 100-inch telescope on Mount Wilson had opened for business. Hubble made good use of his unique opportunity. He took photographs of the Andromeda Nebula (M31). It was the spiral nebula with the largest angular size in the sky (modern photographs trace its diameter at 3°

in the sky—6 times the angular diameter of the Moon.) It was, therefore, a good candidate for the closest spiral nebula. Hubble's photographs resolved it into stars. It was not a gas cloud. It looked like a fuzzy patch because it was made of faint distant stars. He made a sequence of photographs. Some of the brightest stars varied in brightness in a regular way over time. He could recognize them as Cepheid variables—stars whose luminosity varied periodically over periods ranging from days to months (and which were considerably more luminous than RR Lyrae stars). In 1908 Henrietta Leavitt, working at Harvard College Observatory, discovered a relationship between the period of oscillation of a Cepheid variable and its intrinsic luminosity. If you saw a Cepheid variable and measured the timescale of its periodic variation in brightness (in days), you could figure out its intrinsic luminosity (in watts) and therefore determine how far away it was by observing how faint it appeared to be in the sky. But the Cepheid variables Hubble found in the Andromeda nebula were *very* faint and, therefore *very* far away, far outside the Milky Way Galaxy proper. The Andromeda Nebula was so far outside the Milky Way that given its angular diameter, it had to have a physical size of the same order of magnitude as the Milky Way. The Andromeda Nebula was itself a galaxy—just like the Milky Way. The Curtis–Shapley debate had been settled—Shapley was wrong, and Curtis was proved right. People like Immanuel Kant had speculated with good reasons that the spiral nebulae might be entire galaxies like our own, but now we knew. Hubble's evidence settled the case. It would turn out that Hubble's enormous distance estimate for M31 was actually an *underestimate*. But it made the point. M31 was indeed far outside our own galaxy.

The Andromeda Galaxy (M31) is actually 2.5 million light-years away. (I can call it a galaxy now rather than a nebula.) It is a disklike system about 120,000 light-years across. Our galaxy, the Milky Way, is a similar disklike system about 100,000 light-years across. If our galaxy were the size of a standard dinner plate (10 inches across), the Andromeda Galaxy (M31) would be another dinner plate 21 feet away. Light from the nearest star that we see today started on its way about 4 years ago. As you look farther away, you look further back in time. When you look toward the Milky Way's galactic center (in the constellation Sagittarius) 25,000 light-years away, you are seeing it not as it is now but as it was 25,000 years ago, about the time a child left footprints in the

Chauvet Cave in France while viewing the cave paintings there. The light from the Andromeda Galaxy started on its way 2.5 million years ago, when our grandfather species from the genus *Australopithecus* walked the earth.

How can we visualize such large distances? Models can be helpful. At a scale of 1/billion, the entire Earth becomes a marble ½ inch across. A billion is a big number. If Earth is a marble ½ inch across, the Moon is a BB ⅛ inch across, located 15 inches away. Fifteen inches at a 1/billion scale is as far as human astronauts have ever gone—just 15 inches. The Sun, at this scale, is a beach ball 55 inches across, 500 feet away. The nearest star, Proxima Centauri, is a basketball 9 inches across, located 25,000 miles away. That's equal to the entire circumference of Earth, probably more than you drive your car in a year.

Let's shrink things by another factor of a billion, which shrinks Earth to a size smaller than an atom. At this scale of 1/billion-billion, Proxima Centauri is about the size of a hydrogen atom. The typical distance between stars is now about 1.6 inches. Imagine walking in a heavy snowstorm where the falling snowflakes are a less than a couple of inches apart. That is what our region of the Milky Way is like at a scale of 1/billion-billion. Go out at night, and you will see a snowstorm of stars. The Milky Way is about 2.5 miles across at this scale—about the size of a town. When we see the band of the Milky Way, we are seeing the distant lights of our town. Andromeda is another town 64 miles away, at this scale. Hubble was finding more and more galaxies stretching as far as telescopes could see. The typical distances between bright galaxies are about 24 million light-years—140 miles apart on our 1/billion-billion-scale model. Astronomers have seen galaxies as far as 13 billion light-years away—that's 76,000 miles away in our 1/billion-billion-scale model, in which stars are barely larger than atomic size. We are seeing these most distant galaxies as they were 13 billion years ago. This helps us visualize the vastness of the visible universe.

Slipher's Troublesome Redshifts

Meanwhile, Vesto Slipher was observing at Lowell Observatory in Flagstaff, Arizona, taking spectra of galaxies. Slipher found absorption lines in the spectra he took that were at wavelengths associated with particular elements. But he found them shifted slightly from their laboratory values. This is due to the Doppler shift. If a galaxy is moving toward us, succeeding wave crests of light are emitted at locations progressively closer to us as the galaxy approaches, crowding the wave crests together, shortening the wavelength of the light. Blue light has a shorter wavelength than red light, so this causes a "blueshift" in the spectral lines. If the galaxy is moving away from us, each subsequent wave crest reaching us is emitted at locations further and further away, and this stretches out the distance between wave crests. It causes a "redshift" in the spectral lines. The *redshift* z is defined as the fractional shift in the wavelengths of the spectral lines of a galaxy from the laboratory values: $(\lambda_{observed} - \lambda_{lab})/\lambda_{lab}$. For small redshifts (much less than 1), such as Hubble was observing, z is approximately equal to the recessional velocity of the galaxy divided by the speed of light. (In general, the recessional velocity divided by the speed of light is $[(z + 1)^2 - 1]/[(z + 1)^2 + 1]$. As the redshift approaches infinity, the recessional velocity approaches the speed of light, according to Einstein's theory of relativity.) We can experience the Doppler effect with sound waves when a train approaches blowing its whistle. As it approaches, we hear a high-pitch (short-wavelength) sound, and after it passes the pitch becomes lower (longer wavelength) as the train moves away: WHEEEEOOOOOOO.

Doppler shifts had already been used to measure the velocities of individual stars in the solar neighborhood. These velocities were on the order of 20 kilometers per second and showed some stars moving toward us and others moving away—not surprising, if these stars are moving with us on slightly different orbits within our galaxy. When Slipher measured the Andromeda Galaxy, he found a blueshift corresponding to an approach velocity of 300 kilometers per second. This was an enormous velocity, roughly 0.1% the speed of light, much larger than that of individual nearby stars in our galaxy. Today we know that this high approach velocity toward the solar system results in part from the fact that the stars in our solar neighborhood have a mean rotational velocity about the center of our galaxy (about 220 kilometers/second), which happens to be sending us in more or less the direction of where Andromeda happens to be on the sky. Furthermore, the Andromeda galaxy is actually falling toward the Milky Way because of their mutual gravitational attraction. By measuring the differential shift in the spectral lines across the face of an individual galaxy, Slipher was the first to prove that the spiral nebulae were rotating. Slipher continued taking spectra of galaxies and found to his surprise that almost without exception (Andromeda being one) they had redshifts. In general, galaxies were moving away from us. In 1917 he published a table of redshifts. The average recessional velocity of the galaxies in his table was 570 kilometers/second. Astonishing—and unexpected. As he looked at ever fainter (more distant) galaxies, the average recessional velocity had risen from 400 kilometers/second in 1915 to 570 kilometers/second by 1917. The effect was getting more dramatic. On this basis, John Peacock (2013) has argued that Slipher should be given credit for the discovery of the expansion of the universe—because he found so many redshifts, indicating that other galaxies were moving away from us. Peacock has noted that Slipher's redshifts, because they were so large, identified the spiral nebulae as objects outside our galaxy. Slipher did consider the idea that the spiral nebulae might be scattering, but he rejected the notion because they also seemed clustered. Ultimately, Slipher thought galaxies might be moving at high individual velocities edgewise through space, like thrown Frisbees—in that surmise he was incorrect. Slipher deduced that the solar system had an individual velocity of about 700 kilometers/second relative to the mean motion of all the galaxies he saw. The modern value for this is 384 kilometers/second. Slipher's work on redshifts was fundamental and set the stage for what Hubble was able to do later. In addition, as director of the Lowell Observatory, Slipher hired Clyde Tombaugh and started him on the project that would result in the discovery of Pluto. For his contributions, Slipher was awarded the Gold Medal of the Royal Astronomical Society in 1932.

Einstein Has His Say

Theory enters here. In 1915, after 8 years of concentrated effort, Einstein worked out the correct field equation for his theory of general relativity. This theory explained gravity in a revolutionary way, as the result of curved spacetime. Einstein's equation showed how the "stuff" of the universe (matter, energy and pressure) cause spacetime to curve. The right side of the equation describes the mass-energy density and pressure associated with stuff (matter and radiation) at a point. The left side of the equation tells us how spacetime is curved at that point.[1] Particles follow the straightest trajectories they can in the curved space and time. In the same way, a plane follows a great circle trajectory when traveling between two points on Earth. A plane traveling on a straight-line trajectory from New York to Tokyo will pass over northern Alaska. Stretch a string taut between the two cities on a globe to confirm this. That trajectory looks curved on a flat Mercator map of Earth, but on the curved surface of the globe it is the straightest trajectory you can draw—it is called a *geodesic*.

Einstein could then work out the geodesic trajectories of planets. They differed slightly from those Newton had found. The planet Mercury no longer followed the simple elliptical (Keplerian) path that Newton's theory of gravity predicted. The effects were largest close to the Sun. Einstein predicted that the elliptical shape of Mercury's orbit should slowly rotate as Mercury continued to circle the Sun. This rotation of the orbit amounted to 43 seconds of arc per century. A second of arc is 1/3,600 of a degree—so this is a tiny amount of orbital rotation in a century. Nevertheless, astronomers had already measured just this amount of anomalous rotation in Mercury's orbit, an effect that Newton's theory was unable to explain. When Einstein did this calculation, he said he was so excited it gave him palpitations of the heart. When his prediction of 43 seconds of arc per century matched the exact amount of unexplained rotation astronomers had already observed, he was overjoyed.

Einstein then calculated how much light itself should bend as it passed near the limb of the Sun. According to his theory it should be deflected by 1.75 seconds of arc. If particles of light were attracted by gravity in the same way that massive particles were in Newton's theory, the deflection would be only half that—0.87 seconds of arc. The effect could be tested during a total eclipse of the Sun, when the Moon blocks the Sun's bright surface, allowing background stars to be observed near it. In 1919 Sir Arthur Eddington's British expedition took photographs of stars near the Sun during a solar eclipse and compared them with the positions of those stars on photographs taken 6 months later, when Earth had moved so the Sun was nowhere near those stars in the sky. Deflections of 1.98 ± 0.30 and 1.61 ± 0.30 seconds of arc were observed during the eclipse from two different locations, in agreement with Einstein's value within the observational errors (±0.30 seconds of arc) and in clear disagreement with Newton's value. Newton's account of gravity was overthrown. This was the "play of the century" in science. On the day he learned of the eclipse result, Einstein wrote a touching note to his mother, saying: "Good news today." The results were publicly announced at a joint meeting of the Royal Society and the Royal Astronomical Society. J. J. Thomson (the discoverer of the electron) pronounced the result "the most important result obtained in connection with the theory of gravitation since Newton's day." At that moment Einstein moved up to Newtonian stature. The Einsteinian deflection was quickly confirmed by W. W. Campbell and R. Trumpler, in the 1922 solar eclipse; they found a deflection of 1.82 ± 0.20 seconds of arc. Importantly, it has been confirmed with higher and higher accuracy ever since.

In 1916 Karl Schwarzschild found an exact solution to Einstein's equation for a point mass. We know this today as the *black hole solution*. In 1917 Einstein applied his equation to cosmology. He found he could not produce a static solution. Newton had a static universe: stars filling infinite space uniformly, with the gravitational forces on each star on average canceling out. Such a universe could be infinitely old and escape the question of first cause. Einstein in 1917 knew that the individual velocities of stars were small, of order 20 kilometers/second, much less than the velocity of light, and moving randomly. This suggested to him a static universe. So, he added a term to his field equation of General Relativity.[2] The new term was called the *cosmological constant*. It preserved the desirable property of the original equations that energy was conserved locally (in small regions), and the new term was so tiny it did not interfere with the orbit of Mercury or the light bending appreciably. Primarily, it provided a repulsion term that balanced the average gravitational attraction of the stars for each other, allowing a static model of the universe. In his 1917 paper he argued that such a static model was required because of the small velocities of the stars (relative to the speed of light). These, of course, were just stars in our solar neighborhood in the Milky Way. In the Einstein universe, the volume of space was finite and curved but had no edges. It was like the surface of a higher-dimensional sphere called a *3-sphere*. Just as a circle has a finite circumference and a sphere has a finite surface area, Einstein's

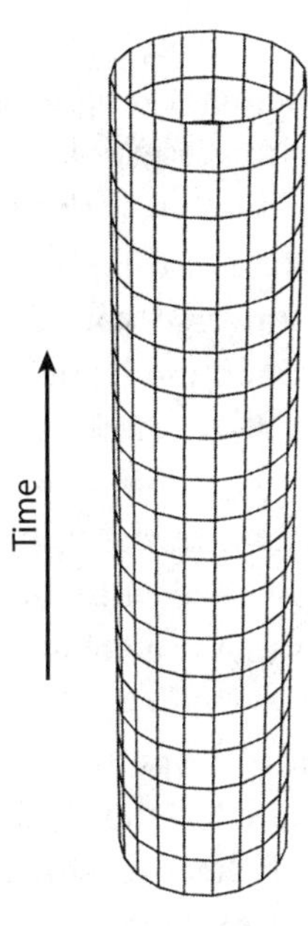

Figure 1.1. Spacetime diagram of the Einstein Static Universe. We are showing only one dimension of space (around the circumference of the cylinder) and the dimension of time (the vertical direction, future toward the top). Worldlines of stars (or galaxies) are straight lines (geodesics) going straight up the cylinder. The circumference represents the circumference of the surface of a higher-dimensional spherical balloon (a 3-sphere) whose radius is unchanging with time. The only real thing is the cylinder itself—the inside and outside have no significance. (Credit: J. Richard Gott, *Time Travel in Einstein's Universe*, Boston: Houghton Mifflin, 2001)

3-sphere universe had a finite volume. If you started off in your spaceship on a voyage in Einstein's static universe and steered as straight a course as you could, you would return to your home planet after having circumnavigated the entire universe—just as you can circle Earth's surface and return to where you started. Because Einstein's universe was static, its volume was unchanging with time. Figure 1.1 shows a spacetime diagram of the Einstein static universe. It shows the dimension of time vertically, with the future toward the top. One dimension of space is shown horizontally. The Einstein static universe spacetime looks like the surface of a cylinder. The static 3-sphere universe is depicted as a circle whose circumference is constant with time. Stack up a bunch of circles (all the same radius) representing the 3-sphere universe at different instants and you get a cylinder. The only thing real in this picture is the cylinder itself. Forget the inside and outside.

That same year, 1917, Dutch mathematician and physicist Willem de Sitter found an exact solution to Einstein's field equation, including Einstein's new cosmological constant term but having no matter in it at all. This was an empty universe—no stars, no galaxies, just empty curved space. (Why should we be interested in such a universe? Because the real universe is of rather low density and so might roughly approximate an empty universe in the limit.) De Sitter's universe looked static as well and seemed to cover only half of a higher-dimensional 3-sphere—as if you had cut Einstein's static universe in half and then thrown half of it away. If you sat at the "north pole" of this universe, you could see down only to its "equator." De Sitter proposed that antipodal points in spherical universe be identified as identical points—as if you had a replica of yourself living at the south pole. That way, the universe you saw (north of the equator) really included all the objects in the universe. The de Sitter universe would thus have half the volume of the comparable Einstein static universe. You would see objects that lived closer to the equator aging in slow motion and emitting light waves in slow motion. If their atoms emitted light at a certain frequency, you would see it at a slower frequency—and a longer wavelength. Thus, in de Sitter space, distant objects showed redshifts. This might explain Slipher's redshifts. These were gravitational redshifts, caused by a photon of light losing energy and increasing in wavelength as it fights its way out of a deep gravitational well. This effect is predicted by general relativity. Clocks on Earth tick a tiny bit slowly relative to those in interstellar space. Einstein, in his work on the photoelectric effect, showed that electromagnetic waves (light) are composed of photons whose energies are inversely proportional to their wavelength. When we send a photon into the sky, it loses energy as it climbs against Earth's gravitational field, and its wavelength increases. (GPS systems using satellites have to take this effect into account when they calculate your position.)

But de Sitter's static coordinate system was incomplete and gave a false impression. It is like the story of a blind man examining an elephant; if he touched only its trunk, he might pronounce that an elephant was most like a fire hose. Figure 1.2 shows our modern understanding of de Sitter spacetime.

It is a 3-sphere universe that starts off with infinite size in the infinite past, contracts to a minimum radius at a "waist" in the center, and then reexpands. It is the gravitationally repulsive effect of the cosmological constant that halts the contraction and causes the subsequent expansion. De Sitter spacetime looks like a corset. The waist appears as a circle. This represents the circumference of the 3-sphere universe at

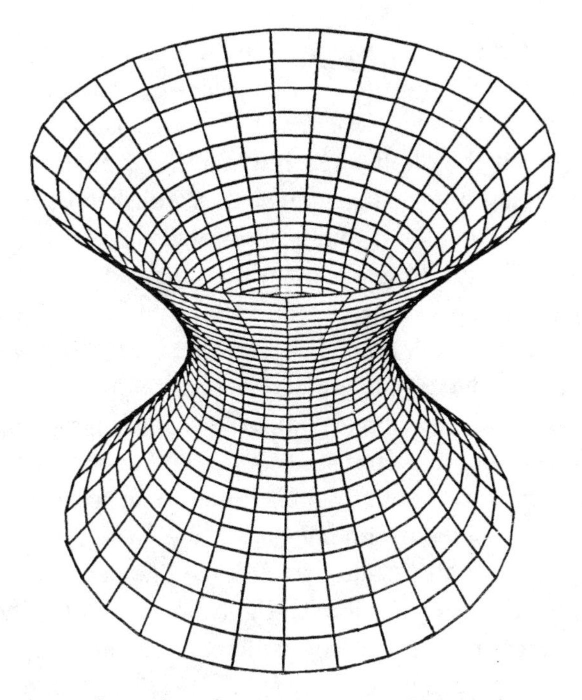

Figure 1.2. Modern understanding of De Sitter spacetime. Diagram shows one dimension of space (around the circumference) and the dimension of time vertically, future toward the top. De Sitter spacetime looks like a corset with a narrow waist at the middle. The horizontal circular cross sections show the size of the 3-sphere universe as it contracts and then expands. The vertical "corset stays" represent possible geodesic *worldlines* (paths through spacetime) of particles. Put a large X through the center of the figure—De Sitter's 1917 coordinate system covered only the left quadrant; leaving out top, bottom, and right quadrants. It made the spacetime look static, but actually it is dynamical, as depicted here. (Credit: J. Richard Gott, *Time Travel in Einstein's Universe*, Boston: Houghton Mifflin, 2001)

the time of its minimum size. The "north pole" is the point at the far left edge of this circle. But the north pole is not just a point that lasts for an instant. It stays around forever. If you live there, your worldline—your path through spacetime—is the vertical corset stay at the far left. De Sitter's static coordinate system, trails after you like a flock of geese following after a leader. The coordinate system covers only the one quadrant of de Sitter spacetime next to the north pole corset stay.

Actually, de Sitter space was a complete spherical universe. It was expanding at late times, which caused distant galaxies to have redshifts. It was a strange universe.

Astronomers at that time were asking themselves if de Sitter space was a useful model for our universe. A piece of Princeton lore (which I heard as a graduate student) was that Henry Norris Russell once asked Harlow Shapley if the Slipher redshifts could be due to the de Sitter effect. No, Shapley reportedly replied, that was impossible, since the globular clusters [Shapley erroneously thought] were at much larger distances than the spiral nebulae, and they showed no redshifts at all! Without proper distances, Shapley found Slipher's redshifts did not even convince him that the spiral nebulae were outside the Milky Way.

In 1922, Alexander Friedmann, in Russia, found an exact solution to Einstein's original equations *without* the cosmological constant. This was a dynamical solution—not static. Like the Einstein static universe, its shape was the surface of a higher-dimensional sphere—a 3-sphere, but its radius changed with time. It started at zero size with a Big Bang. It then expanded rapidly. Galaxies would be like pennies taped to the surface of a balloon: as you blew up the balloon, it got bigger, and the distances between the pennies would increase. If you sat on a penny, all the other pennies would move away from you as the balloon expanded. The space between the galaxies would be expanding. A penny twice as far away would recede from you at twice the velocity.

Start with the balloon at a certain size and consider one penny an inch away from your penny, a second penny 2 inches away, and a third penny 3 inches away. Now slowly blow up the balloon over the course of an hour so that it becomes twice the size. The distances between all pennies will have doubled. The penny that used to be 1 inch away is now 2 inches away—it has been moving away from your penny at a velocity of 1 inch per hour. The second penny that was originally 2 inches away is now 4 inches away: it has moved 2 inches in the hour—moving at a velocity of 2 inches per hour. The third penny that was originally 3 inches away is now 6 inches away—it has moved 3 inches in an hour—a speed of 3 inches per hour. There will be a linear velocity-distance relation. A galaxy 3 times as far away will be receding 3 times as fast.

Like a balloon blowing up, the Friedmann universe is expanding homogeneously as it increases in size after the Big Bang. If you found a linear velocity-distance relation for galaxies like that shown by the pennies, that would support Friedmann's model. More distant galaxies would be receding more rapidly, as did the more distant pennies on the expanding

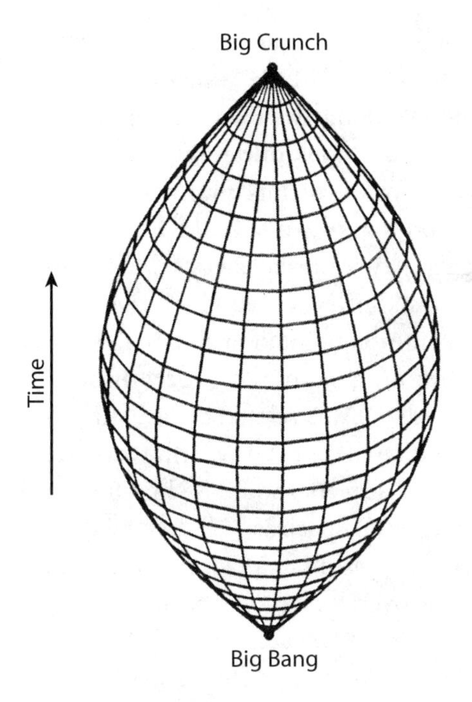

Figure 1.3. Friedmann Big Bang universe of 1922, showing one dimension of space (the circumference of the football shape) and one dimension of time (vertical). Worldlines of galaxies are the vertical seams in the football. The universe is dynamical, starting with a Big Bang at the beginning. The galaxies move apart at first as the circumference of the 3-sphere universe gets larger with time. The gravitational attraction of the galaxies will eventually cause the universe to start contracting, and it will end with a Big Crunch. The only thing real is the "pigskin" itself—the inside and outside of the football have no significance. (Credit: J. Richard Gott, *Time Travel in Einstein's Universe*, Boston: Houghton Mifflin, 2001)

balloon. Friedmann's universe begins in a state of expansion, but eventually the mutual gravitational attraction of the galaxies for each other causes the expansion to slow, finally stop for a moment, and then reverse as the universe begins to shrink. Ultimately, it would collapse again to a point in a *Big Crunch*, when galaxies and stars would be crushed out of existence (see Figure 1.3).

If you had read Friedmann's paper carefully in 1922, you could have predicted that the universe should be seen to be either expanding or contracting. When you observed redshifts of distant galaxies, you could have deduced that you were living in the first half of the universe's lifetime—in the expansion phase. But Friedmann's paper was not widely known. Einstein thought it was a mathematically correct solution to his equation but argued that it was not physically relevant because the low velocities of the stars showed the universe to be static.

Meanwhile, German mathematician Herman Weyl realized that galaxies would not stay at fixed locations in the de Sitter universe, relative to you living at the "north pole." Galaxies would have to accelerate, to fire rockets, to stay at fixed positions in "latitude" in de Sitter space. A gravitational field was pulling them down toward the "equator." Galaxies don't have rockets, so they should be falling away from us in de Sitter space. If galaxies remained at fixed locations in de Sitter space in the static coordinate system centered on you (which is unphysical), they would show a velocity-distance relation that was quadratic: for nearby galaxies that we could observe, a galaxy twice as far away would have four times the redshift. Weyl figured out, however, that if galaxies were launched on free trajectories from a common point of origin in de Sitter space long ago, they would be spreading out now, moving apart through curved spacetime in such a way that redshift was related linearly to distance. A galaxy twice as far away would be moving away with twice the velocity. In that more reasonable case, observers should be looking for a linear velocity-distance relation in de Sitter space as well. In de Sitter's universe you might attribute some of this redshift to a Doppler shift and some to a gravitational redshift, but the overall relation was linear.

Richard C. Tolman at Caltech, in particular, emphasized that one should look for a linear velocity-distance relation in de Sitter space. Various people tried to find a linear velocity-distance relation; they failed to produce convincing results because in addition to spiral nebulae, they were often including globular clusters, which were within the Milky Way and not actually moving away. Hubble knew that if the de Sitter model was correct, he should be looking for a linear velocity-distance relation for galaxies. A galaxy twice as far from us would have twice the redshift.

Hubble Finds the Answer

Hubble used Cepheid variables, brightest stars, and galaxy luminosity as distance indicators to estimate the distances of individual galaxies. He plotted these distances versus redshifts obtained by Slipher and himself, and he found that they followed a linear velocity-distance relation: $v = H_0 d$, where v is the recessional velocity of a galaxy, d is its distance, and H_0 is a constant (today called the *Hubble constant* in his honor). The subscript 0 refers to the fact that it is being evaluated at the present epoch. Hubble found that this linear relation continued all the way out to a recessional velocity of 1,000 kilometers/second. When Hubble published his result in 1929, he mentioned that the effect could be attributed to the de Sitter effect (a combination of the slowing down of atomic vibrations and a general tendency of particles to scatter—Weyl's point). Of course, Friedmann's expanding universe model would predict the exact same linear velocity-distance relation in a far simpler way, but Hubble did not mention that.

Hubble, along with the talented Milton Humason, then went to work on finding distances and redshifts to even more distant galaxies. Hubble wisely subtracted off the peculiar velocity of the Sun. He wanted to obtain the velocities of the other galaxies relative to the center of our galaxy. To go out to large distances he used clusters of galaxies. He knew that galaxies orbiting the center of mass of their cluster could have individual velocities and he understood that he could get a better redshift for the cluster as a whole by taking the average redshift of galaxies in the cluster. For example, he mentioned that the Virgo cluster of galaxies had a mean recessional velocity of 890 kilometers/second, while individual galaxies in the cluster had a total range of 550 kilometers/second about the mean. Hubble and Humason's plot was linear in velocity versus distance out to the most distant cluster, which was receding with a velocity of 20,000 kilometers/second! (See Figure 1.4.) This was spectacular. That cluster was receding at 6.7% the speed of light. It was an astonishing velocity that got people's attention.

Many people date Hubble's discovery of the linear velocity-distance relation to 1929, just as people date the invention of the airplane to the Wright brothers' 1903 flight of 120 feet. But the Wright brothers, like Hubble, had competitors. By the time the Wright brothers flew, Samuel Langley had already flown (unpiloted) model airplanes on flights approaching a mile. Some say Langley should be given credit for inventing the airplane, just as today some argue that Slipher deserves credit for discovering the expansion of the universe. But perhaps more important was the fact that by 1908 the Wright brothers were making flights of more than 123 kilometers. In a similar way, it is the spectacular 1931 paper of Hubble and Humason that really cemented the linear velocity-distance relation and made people take the results seriously. We then tend to trace back retroactively and credit the 1929 paper for the discovery. But it was the 1931 paper that convinced everyone. That's why I chose to show it in Figure 1.4. The entire data set from the earlier 1929

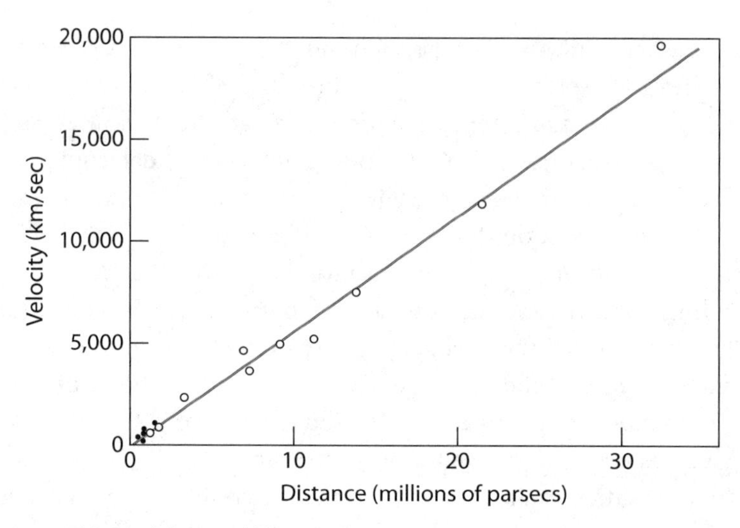

Figure 1.4. This graph from Hubble and Humason in 1931 proved to Einstein that the universe was expanding. Distance is plotted horizontally and recessional velocity is plotted vertically. There is a linear relation: galaxies twice as far away are receding twice as fast. The set of black dots in the lower left corner shows the range of distances and recessional velocities of individual galaxies covered in Hubble's 1929 paper in which he first reported the linear relation. The open circles represent distances and recessional velocities of groups and clusters of galaxies. (Credit: E. Hubble and M. Humason, *Astrophysical Journal*, 74: 43, 1931)

paper is shown by the tiny black data points at the extreme bottom left of the figure.

Hubble's law $v = H_0 d$ implies that galaxies are moving apart today. Velocity multiplied by time gives distance traveled. A galaxy moving with velocity $v = H_0 d$ traveling for a time $t_H = 1/H_0$ will travel a distance d (equal to the distance from us to where it is now). If you play this movie backward in time, all the galaxies will collide at a time approximately $t_H = 1/H_0$ in the past. The time t_H in the past is just enough time to bring a galaxy, which is now at a distance d, back to hit us. Notice that a galaxy will collide with us at a time t_H ago regardless of its distance from us today. This moment where all the galaxies collided in the past is the moment of the Big Bang. At that time, about 13.8 billion years ago, all the galaxies were flung apart from each other like pennies on the surface of an expanding balloon. The space between the galaxies started expanding just as in Friedmann's model. Hubble's law suggested that the universe had a beginning in the finite past. This is extraordinary.

Friedmann's 1922 model, without a cosmological constant, represented a high-density 3-sphere universe, where the gravitational attraction of the galaxies for each other would eventually overcome their kinetic energy of expansion to cause the universe to stop expanding and eventually collapse to zero size in a Big Crunch singularity at the end. You don't want to be around at the Big Crunch! You could determine if this was the type of universe you lived in by comparing the density in the universe you observed today (ρ_0) with a *critical density* calculated from the observed value of the expansion rate: $\rho_{crit} = 3H_0^2/8\pi G$, where G is Newton's gravitation constant and H_0 is Hubble's constant. Given the current best value for the Hubble constant, the value of the critical density is $\rho_{crit} = 8.5 \times 10^{-30}$ gram/cubic centimeter. In Friedmann's models the future of the universe is determined by the value of *omega* (Ω_0) = ρ_0/ρ_{crit}, which is simply the ratio of the density observed today to the critical density. This value was named omega after the last letter in the Greek alphabet, for it was considered to be the last parameter to be sought in cosmology. Once again, the subscript 0 refers to the fact that it is evaluated at the present epoch. If the value of the matter density in the universe today is larger than ρ_{crit}, then $\Omega_0 > 1$, and this corresponds to Friedmann's 1922 model, which starts with a Big Bang and ends by collapsing in a Big Crunch (recall Figure 1.3).

This is a finite universe having positive spatial curvature. It looks like the surface of a balloon but with one higher dimension. As the balloon expands, the volume of space grows. But when the universe starts collapsing in the future, the volume of the universe will shrink as the galaxies crowd together. Eventually, at the Big Crunch, the volume of the universe shrinks to zero as the balloon collapses to zero size. But what if the density in the universe is less than the critical density, and $\Omega_0 < 1$? Friedmann examined this type of model in 1924. In this case, the universe has a negative spatial curvature and looks like the surface of a western saddle, but with one dimension higher (see Figure 1.5).

The saddle extends infinitely, making this a universe with an infinite volume. This model starts with a Big Bang and expands forever. The galaxies weigh so little that their mutual gravitational attraction is unable

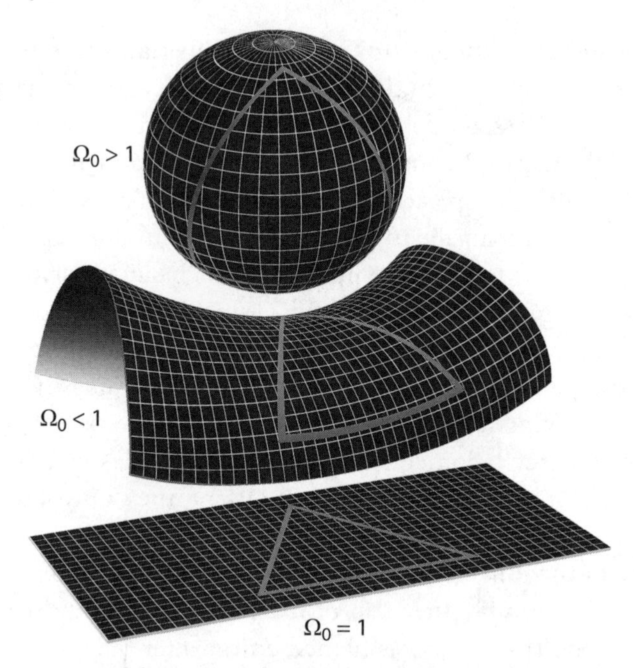

Figure 1.5. Curvature of the universe. If $\Omega_0 > 1$, the universe is positively curved like the surface of a balloon (like the Einstein Static Universe, Figure 1.1, or the Friedmann 1922 universe, Figure 1.3). A positively curved universe is finite with a finite number of galaxies. If $\Omega_0 < 1$, the universe is negatively curved like a saddle. If $\Omega_0 = 1$, the universe has zero curvature, like a flat plane. The last two cases are infinite in extent with an infinite number of galaxies. Triangles have a sum of angles greater than 180° if $\Omega_0 > 1$, less than 180° if $\Omega_0 < 1$, and equal to 180° (the Euclidean value) if $\Omega_0 = 1$. (Credit: NASA)

ever to stop the expansion of the universe. Eventually, Howard P. Robertson (of Princeton and later Caltech) would add a third intermediate case: $\Omega_0 = 1$. Here the universe's spatial geometry is flat like a Euclidean plane. Imagine an infinite, expanding sheet of rubber. The universe at a given epoch has an infinite volume and obeys the laws of Euclidean solid geometry. This third model also starts with a Big Bang and expands forever. It has just barely enough kinetic energy in expansion to overcome the mutual gravitational attraction of the galaxies. Much of the cosmological quest in the twentieth century was centered on the search for the value of Ω_0.

In 1931 Einstein visited Pasadena and met with Hubble. Using contemporary reports, along with Einstein's correspondence and diary,

Harry Nussbaumer has chronicled Einstein's conversion from a static model to a Friedmann Big Bang–type model in 1931. He notes that weighing on Einstein's mind was Arthur Eddington's 1930 demonstration that Einstein's 1917 static model was unstable: perturb it a little and it could start expanding. According to a report by the *Los Angeles Times* on February 5, 1931, Einstein said this in a seminar:

> The redshift of distant nebulae has smashed my old construction like a hammer blow. The only possibility is to start with a static universe lasting a while and then becoming unstable and expansion starting [Eddington's prescription], but no man would believe it. A theory of an expanding universe at the rate figured from the recession of nebulae would give too short a life to the great universe. It would be only 10 thousand million years old, which is altogether too short a time.

Actually 10 thousand million years, or 10 billion years, is close to the true age of the universe—13.8 billion years—but Einstein was worried because at that time, following work by James Jeans, the ages of the stars were thought to be about 6 trillion years. This was based on the plausible but erroneous idea that stars were destroying atomic particles in their cores and converting all of their mass into energy via Einstein's equation $E = mc^2$. (In fact, stars are burning hydrogen into helium in their cores, which releases only 0.7% of their mass in the form of energy. When the nuclear reactions in stars became properly known, the ages of the oldest stars could be calculated accurately and turned out to be of order 13 billion years.) Einstein convinced himself that the age argument might have a loophole—if stars were inhomogeneous in composition—and he became willing to adopt the simple Friedmann models without a cosmological constant. In an April 1931 report to the German Academy, he favored Friedmann's original 1922 spherical $\Omega_0 > 1$, Big Bang–Big Crunch model. By January 1932, he had teamed up with de Sitter to champion an $\Omega_0 = 1$, flat Friedmann-type model that expands forever.

By 1948 physicist George Gamow and his students, Herman and Alpher, were using an expanding $\Omega_0 < 1$ Friedmann model to predict that the early universe should be hot and filled with thermal radiation. The amount of radiation could be calculated from the requirement that the right amounts of hydrogen, helium, and deuterium (heavy hydrogen) got made in the first 3 minutes of the universe's history. As the universe expanded, the wavelengths of this initial thermal radiation would be stretched by the expansion itself and the radiation should have cooled to just a few degrees above absolute zero on the Kelvin scale by now, they calculated. (Kelvin = Celsius − 273. Absolute zero Kelvin represents no molecular motion.) This leftover radiation should be visible as microwaves today, still batting around the universe. Arno Penzias and Robert Wilson discovered this cosmic microwave background radiation in 1965, proving that the universe had a hot, Big Bang beginning. They would win the 1978 Nobel Prize in Physics for this discovery. I worked for them as a graduate student, when their momentous discovery had already become the stuff of legend in astronomy.

There is another, often unsung, hero who should join our cast of characters from the early days: Georges Lemaître. In 1927, he plotted some distance data from Hubble versus redshifts from Slipher and got a linear relation. Lemaître said this meant the universe was expanding. That's before Hubble's 1929 paper. Lemaître published in an obscure Belgian journal and his paper was not immediately noticed. Lemaître was a theorist; he had his own cosmology, which had the same spatial geometry as the Einstein and Friedmann models—the curved surface of a higher-dimensional sphere. It started with a Big Bang and expanded like Friedmann's model did, then stalled at a constant radius for a while, looking like an Einstein static model, and eventually began an accelerated expansion that approximated an expanding de Sitter space at late times. As the matter in the model thinned out in the final phase of expansion, it became quite sparse (nearly empty) and approximated de Sitter space—which Lemaître correctly realized was actually expanding at late times. This whole scenario used Einstein's equations *with* a cosmological constant. Actually, of all the models of that time, it is Lemaître's that comes the closest to what we know today. (We just don't have Lemaître's intermediate phase, in which the universe "coasts" following Einstein's static model.) It can be argued that Lemaître was the first to realize (utilizing data from Slipher and Hubble, of course) that the universe was actually expanding.

Hubble always plotted velocity versus distance, presumably because he thought that the other galaxies were receding from us and, therefore, by implication, that the universe was expanding. But he was rather circumspect on this topic, noting explicitly that the de Sitter solution implied the possibility of other general relativity effects at work—that is, gravitational as well as Doppler redshifts. He didn't want to be wrong.

He wanted credit for discovering the linear redshift-distance law, and he left the interpretation to the theorists.

Later, George Gamow wrote that Einstein once told him that the cosmological constant was "the greatest blunder" of his life.[3] No one much noticed Friedmann's 1922 paper, and Friedmann himself died in 1925, well before Hubble's observations in 1929 and 1931. Friedmann never lived to see his paper vindicated. But if Einstein had written the Friedmann paper instead, *everyone* would have noticed. If Einstein had not invented the cosmological constant, his theory would have predicted that the universe should be expanding (or contracting) in *advance* of Hubble's observations. Then, when Hubble's observations came out, Einstein's theory of general relativity would have been proven correct in extraordinary fashion. Predicting ahead of time the expansion (or contraction) of the universe and having it confirmed by Hubble would have been a much more spectacular confirmation of general relativity than a little light bending around the Sun.

But in a further twist, in 1997 astronomers discovered that the expansion of the universe is actually *accelerating* today. This would not occur in a Friedmann-type model. Remarkably, it requires something that looks exactly like the cosmological constant term Einstein proposed. Einstein has triumphed after all, and so has Lemaître, whose cosmological model ended with an accelerating period of expansion, just as we now observe. Einstein's invention of the cosmological constant had not been a blunder after all.

Before Hubble, the universe consisted of the Milky Way. By the time he was finished, Hubble had left us with a universe filled with a myriad of galaxies—spirals, ellipticals, and irregulars—and these galaxies were rushing away from us. The further away they were, the faster they were receding. Hubble left us with an expanding universe. In his 1931 paper, Hubble noted that the proposed 200-inch telescope on Palomar Mountain could see galaxies out to a redshift of 60,000 kilometers/second. He was the first to observe with the new telescope when it opened in 1947. But he didn't get to use it as much as he had hoped. He died in 1953. Yet the Hubble Space Telescope, named in his honor, has now observed galaxies at a distance of 13 billion light-years, receding from us at nearly the speed of light. One hundred forty billion galaxies lie within the range of this telescope. This is Hubble's universe.

Chapter 2

Zwicky, Clusters of Galaxies, and the Discovery of Dark Matter

I knew Fritz Zwicky when I was a young postdoc at Caltech. Whenever Zwicky was spoken of, the word *genius* would hover about. It was not a word people used lightly. There were two people whom visiting colloquium speakers had good reason to fear when they came to Caltech: Zwicky and Richard Feynman. They usually asked questions or made comments during the weekly physics colloquiums. A typical Feynman question might be: "But doesn't that violate energy conservation?" It was usually difficult to recover from such a question. Zwicky's typical comment would be: "I did this already in 1933!" and he would give the reference. You would go to the library (back in those days) and look it up, and sure enough, he had written a seminal paper on that topic.

For one thing, Zwicky proposed the existence of neutron stars. When a star had exhausted the nuclear fuel in its core, its core would collapse, and this process would throw off the star's outer layers. If it were a star like the Sun, the core would end up being a white dwarf star about the size of Earth. This was held up against collapse by the fact that because of quantum mechanics, electrons do not like to be squeezed together. If the core of the star were more massive than 1.4 solar masses, the white dwarf state would be unstable to further collapse and the electrons would combine with the protons to form neutrons. Without the presence of the low-mass electrons, the pressure holding the star up against gravity drops and a neutron star only about 22 kilometers across will form. This is basically a giant atomic nucleus, stabilized by gravity. It is held up against further collapse by the fact that neutrons do not like to be squeezed together either.

Zwicky and Walter Baade proposed the existence of such neutron stars in 1933. They argued that the energy generated from the collapse to form a neutron-star core could power a supernova explosion. Jocelyn Bell would discover neutron stars, called *pulsars*, in 1967. She found them by observing periodic radio pulses. As neutron stars rapidly rotated, their magnetic poles swinging around produced a rotating radio lighthouse beam, which swept by Earth in a periodic fashion. During World War II, Zwicky worked on jet engines, developing the jet-assisted takeoff (JATO) engine for planes on short runways. Truman awarded him the Presidential Medal of Freedom for this work. In 1957, shortly after *Sputnik*, he exploded charges in an Aerobee rocket at high altitude, blasting pellets into the upper atmosphere. These were observed from Mount Wilson Observatory as artificial meteors. It is thought that one of them may have been the first manufactured object to have achieved escape velocity from Earth. He discovered 122 supernovae, a record at the time. He was interested in compact galaxies and noted that they might be mistaken for nearby stars. In this he anticipated *quasars*—bright galactic nuclei which were indeed mistaken at first for stars in our own galaxy. It was Maarten Schmidt, also at Caltech, who discovered that these quasars were rapidly receding from us due to the overall expansion of the universe and were, therefore, according to Hubble's law, very distant, highly luminous objects. Quasars are now understood to be powered by hot gas spiraling inward toward supermassive black holes in galactic nuclei.

Zwicky's office at Caltech was in the subbasement of Robinson Hall, which housed the astronomy department. This building had so many basement levels because it had originally included a solar telescope. On top of the building were coelostat mirrors in an observatory dome, which directed the Sun's image to the basement. Zwicky's office was in the subbasement where the graduate students' offices were located. Zwicky was very popular with the graduate students, who regarded him as a comrade-in-arms. The graduate students used to say that Robinson Hall was like an ocean liner: the professors (whose offices were on the second floor) were up on the promenade deck, while all the work got done down in the engine room (the basement). Zwicky once asked me what I was working on. I said I was working on the formation of spiral and elliptical galaxies. He said, "I can take a picture of a spiral galaxy and make it look like an elliptical"—a typically wry and skeptical response—something that students and postdocs such as myself admired.

Some of Zwicky's most important work was on clusters of galaxies. In 1933 he made an estimate of the mass of the Coma Cluster of galaxies. Astronomers measure mass by using Isaac Newton's formula for orbiting bodies: $v^2 = GM/R$. If you plug in the orbital velocity of Earth, $v = 30$ kilometers/second, and the radius of Earth's orbit, $R = 93$ million miles, and take the known value of Newton's gravitational constant G, you can find the Sun's mass, $M_{sun} = 2 \times 10^{33}$ grams. If you plug in the orbital velocity of our Sun about the center of our galaxy, $v = 220$ kilometers/second, and the radius of the Sun's orbit about the center of the galaxy, $R = 25{,}000$ light-years, then you can find the mass of our galaxy interior to the orbit of the Sun, about 60 billion solar masses. So astronomers knew how much galaxies weighed—at least their luminous parts. When Zwicky plugged the orbital velocities of the galaxies he had found in the Coma cluster (about 1,700 kilometers/second) into the formula, along with the radius of the cluster, he obtained a total mass for the cluster (about 10^{15} solar masses in all) that was about 10 times larger than could be accounted for by the combined mass in the individual luminous galaxies. Newton's orbital formula could be safely applied only if the cluster was stable: neither expanding nor contracting. Zwicky observed that the cluster had a stable spherical shape and was centrally condensed. It looked rather like a globular cluster of stars, except that it was a cluster of galaxies. Figure 2.1 shows the map Zwicky made of the positions of the galaxies in the Coma cluster on the sky.

It had spiral galaxies on its outskirts but mostly elliptical and S0 galaxies near the center. An *elliptical galaxy* is elliptical in shape with old stars and no gas. An *S0 galaxy* is a spiral of type 0, meaning that it has an elliptical bulge in the center consisting of old stars like an elliptical, plus a dinner-plate disk of stars extending out from that. This disk of stars has no gas and also consists of old stars. *Spiral galaxies* are like our own—a central elliptical bulge of old stars with a disk of younger stars and gas extending from it. This disk shows spiral arms like a pinwheel. Ellipticals and S0 galaxies are called *early-type galaxies*, since they have

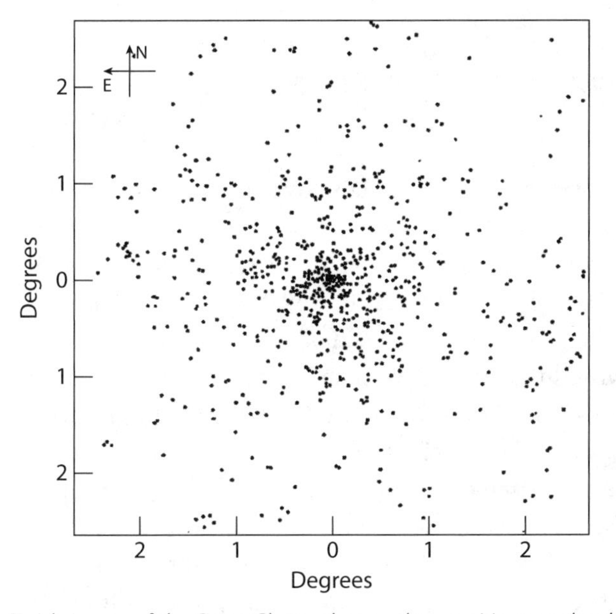

Figure 2.1. Zwicky's map of the Coma Cluster shows galaxy positions on the sky as dots. (Credit: F. Zwicky, *Astrophysical Journal*, 86: 217, 1937)

old stars only, whereas spiral galaxies (and irregular galaxies like the Magellanic Clouds) are called *late-type galaxies* and have young, bright blue stars as well as old stars and show active star formation still occurring. Thus, the Coma Cluster had early-type galaxies in its central regions but late-type galaxies in its outskirts.

In the very center of the Coma Cluster, two giant elliptical galaxies orbited each other like wary heavyweight boxers circling each other in the ring. One of the two giant ellipticals had numerous dwarf galaxies accompanying it. These two big ellipticals had perhaps grown large by gobbling smaller galaxies. Importantly, Zwicky found that the *crossing time* of the cluster—the time it would take a galaxy to move from one side of the cluster to the other—was much shorter than the age of the universe. If the cluster were *not* bound by the force of gravity as Newton's formula implied, the cluster would quickly disperse. It looked like an equilibrium structure and so the galaxies must be orbiting. Holding galaxies in those orbits would require lots of mass, more mass than the luminous galaxies could account for. Zwicky correctly concluded that most of the mass must be what he called *dark matter*, that is, nonluminous matter lying between the galaxies.

Confirmation of Zwicky's Discovery of Dark Matter

Zwicky's astute research lay dormant for many years. Then, in the 1970s, Jeremiah Ostriker and Jim Peebles (of Princeton University) discovered that our galaxy, if solely composed of its known stars, would be unstable. They proposed that it must be stabilized by a spherical halo of invisible dark matter. With Amos Yahil, they collected data on individual galaxies, small groups of galaxies (having only a few members), and clusters of galaxies and argued that the mass of galaxies went up linearly with radius—the further out you went, the more mass you saw. If the density of dark matter fell off like 1 over the square of the radius, it would all add up so that the total mass of the halo inside a given radius would grow like the radius. This would mean that according to Newton's laws of gravity, the rotational velocity of stars around the galaxy would be *constant* with radius, rather than falling off with the square root of the radius, as occurs for planets in the solar system (where most of the mass is concentrated in the Sun). Einasto, Kaasik, and Saar came to similar conclusions in a 1974 paper by studying the velocities of companion galaxies.

If the mass of a galaxy were primarily concentrated in the luminous stars we observe, when we got beyond the optical boundary of the galaxy (the matter we could see), we should observe the orbital velocities of gas orbiting the galaxy drop as one moved further away. Ken Freeman (in 1970) had already noted that rotation curves were not falling off in this way, implying large amounts of undetected matter at large radii. In 1975 Morton Roberts and his colleagues measured hydrogen gas orbiting in the plane of the Andromeda galaxy out to about three times the optical diameter of the galaxy and found the rotational velocity to be unchanging with radius. That is exactly what one should observe if the galaxy's total mass was going up like the radius. But the luminosity contained inside a radius R was not going up proportional to R—evidence that an invisible dark matter halo must extend several times as far as the extent of the stars. This effect was soon demonstrated in great detail by Vera Rubin, confirming halos of dark matter in some five dozen galaxies.

Determining the amount of dark matter associated with galaxies was crucial to getting a better estimate of the total matter in the universe, which in turn was crucial to deciding the basic cosmological question of the fate of the universe. If the total mass density in the universe today were greater than a critical value, the universe would eventually recollapse in the future as in Friedmann's original model from 1922. By examining groups and clusters of galaxies, Ostriker, Peebles, and Yahil estimated that the total amount of matter in the universe is about 20% of the critical density required to eventually halt the universe's expansion in the future. Independently, Einasto, Kaasik, and Saar (1974) also estimated 20%, based on velocities of companion galaxies. This suggested that the universe would continue to expand forever. Both estimates suggested $\Omega_m = 0.2$ and a Friedmann universe that would expand forever. The amount of dark matter they found was consistent with what Zwicky had found and also with the amount required to explain why our galaxy and Andromeda were not participating in the general expansion of the universe but instead were falling back toward each other. Dark matter would have to be boosting the mass of both galaxies for this to occur. Dark matter was needed to explain the stability of galaxies, the rotation rates in their outskirts, and the masses of groups and clusters of galaxies. But what could this dark matter be? Could it be intergalactic gas or black holes, or even planet-sized objects—jupiters that did not glow appreciably? Or was it something as yet unknown?

Gravitational Lensing

Once again, Zwicky had pioneered another important tool for investigating dark matter—gravitational lensing. Einstein's theory of general relativity predicted that light would be bent traveling through curved spacetime. This had been observed for starlight passing near the Sun during the solar eclipse of 1919. In about 1936 Einstein received a letter from an amateur astronomer suggesting that if light could be bent by a massive object, then couldn't light from a distant star lined up directly behind another star be lensed to show up as a bright ring circling the foreground star in the sky. Einstein figured out that, in the typical case, where the two stars were not *perfectly* lined up, the background star would appear as two images, one on each side of the foreground star in the sky. Light coming from the distant star would pass by the foreground star on one side and be bent left to intersect Earth, while light passing the foreground star on the other side would be bent right to intersect Earth as well. Gravity would create a cosmic mirage—a double image of the background star.

Einstein thought the chances of this being observable were low considering that the double image of the background star would be such a close double image in the sky that our telescopes could not resolve it as a double. Henry Norris Russell of Princeton read Einstein's paper and immediately remarked that one could actually observe such an event because as the background star passed close enough behind the foreground lensing star, the sum of the two lensed images of the background star would be brighter than its original unlensed image. Thus, as the background star passed behind the foreground star, it would show a characteristic brightening.

In 1937 when Zwicky heard about this, he immediately applied the magnification and double-image ideas to galaxies. He knew what clusters of galaxies weighed and realized that this made them excellent gravitational lenses: a nearby cluster of galaxies might act as a lens to brighten and magnify a distant galaxy behind it due to the influence of dark matter. In this way a cluster of galaxies could act as a telescope to further magnify our view of even more distant galaxies, ones that would normally be too faint to see. Again, Zwicky was decades ahead of his time.

In 1979, Walsh, Carswell, and Weymann discovered the double quasar 0957 with two images in the sky, one on each side of a foreground galaxy. The two quasar images were about 6 seconds of arc apart in the sky, and the lensing galaxy was seen between them. The lensing galaxy was bending the light from the distant quasar, creating two images on opposite sides of the galaxy, as shown in Figure 2.2.

In 1997, I would observe this system as part of a team led by Tomislav Kundić and Ed Turner. We found a sudden change in the luminosity of one of the two distant-quasar images, followed 417 days later by an equal change in the luminosity of the other one. The distant quasar was about 8.9 billion light-years away. The light in the two images journeyed to us on two separate paths in curved spacetime, as shown in Figure 2.2.

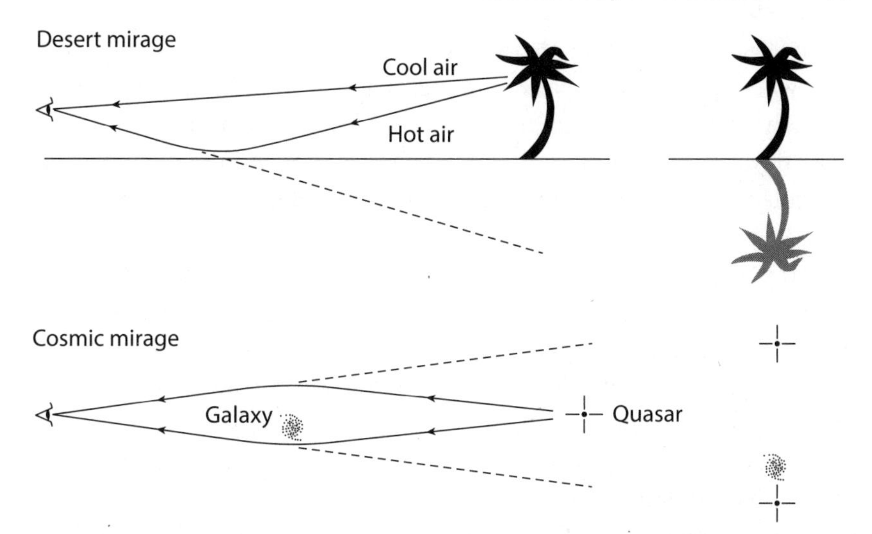

Figure 2.2. Gravitational lensing produces a cosmic mirage by bending light, just as a desert mirage is produced by bending of light in the air due to temperature differentials. In the desert mirage we see images of a distant palm tree and its "reflection" in what seems like a lake of water. Gravity bends light, as Einstein showed, because light is traveling in curved spacetime. Here we see two images of a distant quasar on opposite sides of a lensing galaxy. (Credit: J. Richard Gott, *American Scientist*, 71: 150, 1983)

These two paths had lengths that differed slightly—by 417 light-days. That accounted for the time delay. If the distant quasar had been twice as far away, the time delay would have been twice as great. By modeling the system and using the observed time delay, we could determine the distance to the distant quasar, and that distance, combined with its redshift, allowed us to determine the Hubble constant. We found $H_0 = 64 \pm 13$ kilometers/second/megaparsec (where 1 megaparsec = 3.26 million light-years). This came from just one direct measurement and did not depend on any intermediate steps in the distance ladder, such as the distances to Cepheid variable stars. This agrees with our best current estimate of $H_0 = 67 \pm 1$ kilometer/second/megaparsec from the Planck Satellite collaboration and from the Sloan Digital Sky Survey using a variety of data, including data from the cosmic microwave background. Plugging this current estimate into Hubble's Law, $v = H_0 d$, means that if a galaxy is 10 megaparsecs (32.6 million light-years) away, on average, we will find it to be receding from us at a velocity of about 670 kilometers/second.

In 1981 I wrote a paper saying how quasar 0957 could be used to test whether dark matter was made of jupiters—intergalactic planets with the mass of our Jupiter. These would not glow and would, therefore, be dark. If these constituted the dark matter, then the gravitational microlensing caused by all these jupiters would cause twinkling of the quasar images on a timescale of about a year. Each image would have different jupiters passing in front of its light beam as the jupiters orbited in the lensing galaxy, and so the twinkling in the two images would be independent and uncorrelated. But when our group observed 0957, we found that the two images were changing in synchrony with a time delay. Only an intrinsic variation of the quasar would show up with a time delay in the two images. We saw no twinkling. The dark matter was not made up of jupiters.

Meanwhile, Bohdan Paczyński and Charles Alcock had formed two independent groups to look for jupiters in our own galaxy's heavy halo of dark matter. These jupiters were called MACHOS (massive compact halo objects). One could look for gravitational-lensing twinkling of stars in either the Magellanic Clouds or in the central bulge of our own galaxy. Since the lenses were so close to us, one needed to inspect millions of background stars to find a twinkle. But with modern digital cameras and computer software, one *can* inspect millions of stars. In the end, gravitational microlensing events were found, but they were traced to ordinary stars, and not enough events were discovered to explain the mass of the heavy halo that we knew was there from orbital dynamics. Therefore, the dark matter in our own galaxy was apparently not made of jupiters. Still, the technique was accurate enough to detect planets at distances of tens of thousands of light-years. Occasionally, when an intermediate-distance ordinary star was passing near enough to the line of sight to a distant star to cause a microlensing brightening, an *additional* brightening of the background star would occur briefly because a planet orbiting the lensing star was also crossing the path of the light beam from the distant star at the same time. This would cause a minitwinkle within a twinkle. In this way a number of Neptune-mass planets have been discovered orbiting distant stars.

In 2012, Paul Schechter, Joachim Wambsganss, and their colleagues David Pooley, Saul Rappaport, and Jeffrey Blackburne examined microlensing in quadruple lenses. When the lensing galaxy (or cluster) is directly in front of the quasar to be lensed, it tries to form a perfect ring around the lensing galaxy in the sky. But if the distribution of mass in the lensing galaxy is slightly elliptical instead of perfectly spherical, as is typically the case, instead of a bright ring one gets four bright images of the distant quasar. These four images are arranged roughly in a ring around the lensing galaxy in the sky. From the angular positions of these lensed images around the ring, one can predict the relative brightnesses of the four lensed images if the mass distribution is smooth. The time delays between the arrival of the images are short in this case (a few days at most), so one can correct for the intrinsic brightness variations of the background quasar. If the observed brightnesses of the four lensed images do not agree with those expected from the smooth model, the errors must be due to microlensing by compact objects such as jupiters, stars, or black holes along the line of sight. By careful study of the microlensing in 14 quadruple-image lens systems, they were able to conclude that the material along the lines of sight through the galaxy to the quasars was 93% in a smooth component and 7% in compact objects: that is, stars, jupiters, or black holes (from stellar mass black holes to supermassive black holes). The amount of microlensing seen was just that expected from normal stars; thus most of the mass must be in the form of a dark smooth component. This accords with the picture in which dark matter is a smooth component—it's not made up of jupiters, neutron stars, or black holes.

In 1987, Paczyński recognized that giant arcs seen in clusters of galaxies were actually gravitationally lensed images of distant background galaxies, just greatly magnified and distorted—exactly as Zwicky had foretold! (See Color Plate 2.)

What Is Dark Matter?

We know by studies of microlensing that dark matter is not made of compact objects like stars or planets or black holes. But could it be in smoothly spread intergalactic gas? Consider the Bullet Cluster studied by Markevitch and colleagues in 2004 and shown in Color Plate 3. These are two spherical clusters of galaxies that have collided and passed through each other. As the clusters collide, most of their galaxies will miss each other. Even if two galaxies hit, they also pass right through each other. When you look up into the night sky, you see thousands of stars, but most lines of sight are black, passing through empty space. The galaxies are likewise mostly empty space, and their sparse star fields pass right through each other. In the Bullet Cluster, although the two clusters have passed through each other, each cluster of galaxies also had diffuse intergalactic and interstellar gas, and this *does* collide. We can see a double ball of hot gas in the space right between the two clusters today, glowing brightly in the X-rays (shown in red as mapped by the Chandra X-ray Observatory satellite). If the gas wasn't hot before, it is heated by the collision so that we can see it in the X-rays. Gravitational lensing distortion of background galaxies allows astronomers to map the distribution of dark matter in this system (shown in blue). Where is the dark matter located? Is it in a double lump in the center, like the gas, as would be the case if the dark matter were made of intergalactic gas? No. It is in two widely separated lumps (blue), centered on where the two clusters of galaxies are now. This means that the dark matter could not be made out of ordinary matter in the form of gas.

Low-mass objects like planets or black holes would pass through like the galaxies, but we have already ruled these out. Dark matter could be made of *weakly interacting massive particles* (WIMPs), as proposed by Princeton physicist Jim Peebles. These would be exotic elementary particles, more massive than the proton. They would interact weakly, neither glowing nor significantly interacting with photons or particles of ordinary matter except by gravitation, which affects all particles through the curvature of space. If the dark matter were made of WIMPs, the distribution of mass within galaxy halos would be smooth, consistent with microlensing studies, and dark matter would pass through just as the galaxies do, to end up located in two widely separated lumps just as we see in the Bullet Cluster.

As we shall see in Chapter 5, if dark matter is made of WIMPs, it can also explain the *amount* of galaxy clustering we see, given constraints imposed by the cosmic microwave background. If the dark matter were made of ordinary matter, it would have started gravitationally clustering later and not had sufficient opportunity to grow up into the structures we observe today.

There is another reason for believing that the dark matter is not made of ordinary matter: if all the dark matter were ordinary matter (or black holes made of ordinary matter that had collapsed or jupiters made of ordinary matter), nucleosynthesis in the early universe would have left us today with an amount of deuterium (heavy hydrogen) less than what we observe. Deuterium is fragile and is burned up rather than made in stars, so you can't make it later. You need to make the correct amount in the early universe. If the dark matter was made of WIMPS, these WIMPS would not increase the density of ordinary matter undergoing nucleosynthesis in the early universe, and we would predict the correct amount of deuterium we observe today.

Thus, a number of lines of evidence point to exotic elementary particles as the source of dark matter. Study of the strength of temperature fluctuations in the cosmic microwave background as a function of angular scale by the Planck satellite now allows us to calculate that the amount of ordinary matter corresponds to $\Omega_{\text{ordinary matter}} = 0.048$ (in agreement with nucleosynthesis results), while the weakly interacting dark matter component is $\Omega_{\text{dark matter}} = 0.260$, making the total matter content of the universe add up to $\Omega_{\text{m}} = 0.308$ (close to the early rough estimates of 0.2).

The search is on to find the particles responsible for dark matter. WIMPs could be *supersymmetric* partners of known particles. (The theory of *supersymmetry* says that every known elementary particle type has a partner particle.) The Large Hadron Collider is searching for such particles. It recently discovered the Higgs boson, so hopes are high. Another way to look for dark matter particles is to look for collisions of these particles with nuclei of supercooled argon. Such experiments are located deep in mines and surrounded by scintillation detectors to rule out collisions by normal particles. If a supersymmetric partner for a known particle is discovered, it would support the theory of supersymmetry: the electron has the *selectron*, quarks have *squarks*, the photon has the *photino*, and so forth. Superstring theory, which is our current best candidate for a "theory of everything," predicts that such supersymmetric partners should exist. The dark matter might be made of the lightest supersymmetric partner; a *neutralino* (partner of a neutrino) is one possibility, and the *gravitino* (partner of the graviton) is another.

Another possibility for dark matter are *axions*, a different kind of elementary particle. These are light particles that act in concert to behave like dark matter. A third possibility are *Kaluza-Klein particles*, particles associated with extra, unobservable, microscopic spatial dimensions that the universe may possess, according to superstring theory. Today the study of dark matter, which Zwicky initiated, leads us right to the forefront of particle physics.

Chapter 3

How Clusters Form and Grow—Meatballs in Space

I met Jim Gunn when I was a first-year graduate student at Princeton. I already knew of him by reputation. He had been a star graduate student at Caltech and had come to Princeton as a new assistant professor. If you want to visualize him, think of a young George Lucas at the time he was directing his first *Star Wars* movie. Like me, Jim had been an amateur astronomer as a teenager, but he had built his own telescope, which had been covered in *Sky and Telescope* magazine. At Caltech he had codiscovered what came to be known as the Gunn-Peterson effect: a method to detect intergalactic hydrogen by observing a broad absorption band in the spectra of highly redshifted quasars.

Princeton's astrophysics department has a system where graduate students work with different professors on three different research projects in their first 2 years. I was assigned to work with Jim Gunn, and I was quite excited to see what project he would have in mind. He wanted to study the formation and growth of clusters of galaxies. It would be my job to work out the formulas. Jim felt that after clusters had formed, further material would fall in and this process could be used to measure and test for the presence of intergalactic gas.

Jim and I were going to be working in the context of the standard hot Big Bang model founded by George Gamow and his students Alpher and Herman. They had predicted that if the universe started in a Big Bang, at early times it should have been hot—filled with hot thermal radiation. Compressing gas as one traced backward in time toward the beginning would make it hotter and hotter. At about 1 second after the beginning, the temperature would be about 10 billion degrees. As the universe expanded, the thermal radiation would redshift to longer wavelengths due to the stretching of space and therefore cool. At about 3 minutes after the Big Bang, the temperature would drop enough for protons and neutrons to start fusing into nuclei. This nuclear burning could create the deuterium, helium, and lithium that we observe in the universe today. More helium and heavier elements could be built later in stars.

As the universe expanded by a factor of 10, the radiation would become 10 times longer in wavelength and 1⁄10 the temperature. Eventually, at about 380,000 years after the Big Bang, the temperature of this thermal radiation permeating the universe would drop to about 3,000 degrees on the Kelvin scale (3,000 K)—cool enough for electrons to combine with protons to make neutral hydrogen. Before that epoch, the electrons, which carry electric charges and are independently moving around, would have been coupled closely to the thermal photons and the electrons would drag the protons with them. Thermal radiation has a large radiation pressure and is stiff—it resists being squeezed, thereby preventing any fluctuations in the density of ordinary matter from growing. But after the universe has become neutral—after the negatively charged electrons have combined with the positively charged protons to create neutral hydrogen (with zero total charge)—the neutral atoms are free to move relative to the thermal radiation. Astronomers call this epoch *recombination*, even though the electrons are really *combining* with the protons for the first time. After this epoch, the universe becomes transparent, and photons from that epoch can fly freely to us today. We should be able to see them.

After recombination, the universe expands a further factor of 1,090 to reach the present epoch. The thermal radiation cools off to a temperature of 2.725 K, just slightly above absolute zero on the Kelvin scale. (This is quite cold—100 times colder than the temperature of ice cubes melting in your drink.) The wavelengths of the photons in this thermal radiation have by the present epoch redshifted (stretched) until they are in the microwave region of the electromagnetic spectrum. This radiation fills the universe today and comes at us from all directions. In

1948, Herman and Alpher had predicted its current temperature to be 5 K, based on calculations of the formation of light elements in the early universe and comparison with the observed amounts of these elements seen today. Penzias and Wilson discovered this radiation in 1965. Its temperature (2.725 K) is very close to Herman and Alpher's original estimate. We call this radiation the *cosmic microwave background radiation*. When we see this radiation, which comes to us from all over the sky, we are looking out in space 13.8 billion light-years and back in time 13.8 billion years, to just 380,000 years after the Big Bang.[1] That is the last time the photons we are seeing today had been scattered by electrons before coming directly to us during the universe's transparent era.

For our research project, Jim Gunn and I started with a uniformly expanding universe at recombination, containing a spherical region with a slight density enhancement built in. This would be the seed from which a cluster of galaxies would grow. We had a very simple model: inside a spherical volume in the early universe the density was above average, while beyond that sphere the universe was of exactly average density. It was a lone cluster trying to form in an otherwise average-density universe. The sphere would start off with an expansion rate determined by Hubble's law, just as for the rest of the universe. But the extra mass it had inside would cause it to decelerate more than the universe as a whole, and so, after a while, the radius of the sphere would have expanded by a factor that was *less* than the factor by which the universe as a whole had expanded. Let ρ be the average density in the universe and $\delta\rho$ be the excess density (above the average) within the sphere. Because the sphere is not expanding as fast as the universe due to the action of gravity, the *fractional density enhancement* inside the sphere relative to the rest of the universe, $\delta\rho/\rho$, will become larger than it was originally. Eventually, the extra gravity inside the sphere would stop the expansion of the sphere and it would reach a maximum radius, R_{max}, as it stopped expanding. Once the sphere stops expanding, it begins to collapse under the influence of gravity.

By this time, galaxies have already formed (a process I will discuss later). If the cluster were perfectly uniform, it would collapse to a point within a time equal to the time it took to expand out to its maximum radius. But it is not perfectly uniform, and, therefore, the galaxies will miss each other as they fall through the center. We called the time from the Big Bang to the completion of the collapse of the cluster the *collapse time,* T_c. The cluster starts off expanding like the rest of the universe but slows its expansion due to the extra gravitational attraction of the excess mass inside and eventually stops expanding at a radius R_{max} at a time after the Big Bang of ½T_c. It takes an equal time ½T_c to collapse into the center. After the point of collapse, the galaxies pass through the center in a disordered way and expand back out; the distribution sloshes around for a while and quickly reaches equilibrium. This occurs at a time $\frac{3}{2}T_c$ after the Big Bang. At this point, the random velocities of the galaxies are resisting the gravitational attraction of the mass of the cluster, as the galaxies orbit within the cluster. The equilibrium radius of the cluster is about ½R_{max}. Because the volume of the cluster is 8 times less than before, the equilibrium cluster is about 8 times as dense as it was when it reached maximum expansion with a radius of R_{max}. See Figure 3.1.

Meanwhile the universe has been continuing to expand since the moment when the cluster reached a radius of R_{max}. In a high-density universe that has just the critical density, so that $\Omega_0 = \Omega_m = 1$, we found the cluster was 5.5 times denser than the universe as a whole when the cluster turned around; by $t = \frac{3}{2}T_c$ the cluster has grown 8 times denser, while the universe has thinned out by a factor of 9 due to its continuing expansion. That makes the just-formed equilibrium cluster $5.5 \times 8 \times 9 = 396$ times as dense as the universe as a whole. After that, the cluster stays the same size and the same density (it's in equilibrium), while the universe continues to expand, its density continuing to go down with time. Thus, if we see a relaxed cluster in equilibrium, we expect it to be *at least* 396 times as dense as the universe as a whole.

This is the case for a high-density universe with $\Omega_m = 1$. If the universe has a lower density than this (if $\Omega_m < 1$), the density enhancement of the cluster over the background is larger than the background by a factor of *at least* $396/\Omega_m$. This is how clusters formed after the Big Bang. Only a small positive density fluctuation was needed at recombination. The operation of gravity inexorably turns it into a large density enhancement that is easy to see later. We found that the collapse time of the cluster T_c was approximately inversely proportional to the initial fractional density enhancement $\delta\rho/\rho$ raised to the $\frac{3}{2}$ power (written as T_c proportional to $[\delta\rho/\rho]^{-3/2}$). A bigger initial fractional density excess will cause a shorter collapse time T_c.

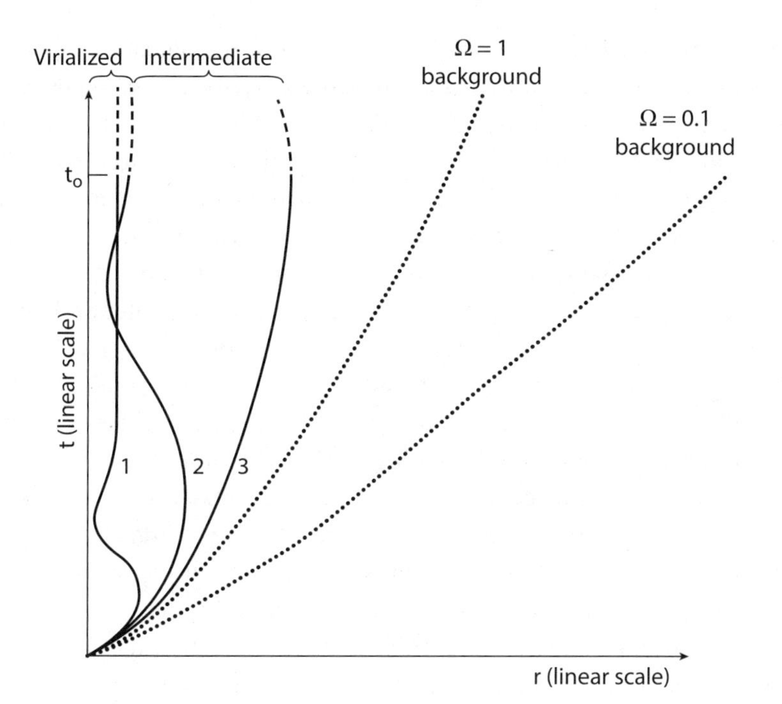

Figure 3.1. Cluster collapse and infall. Time is vertical. The radius of the cluster proper is shown expanding with the universe at first (1), then collapsing and relaxing (this is called *virialization*) to constant size. Shells of matter beyond the cluster proper are shown infalling later (2) to add to the cluster's mass. Shells of matter further out (3) are still infalling today (at time t_0). The overall background expansion of the universe is also plotted for two cases: a $\Omega = 1$ Friedmann universe of just-critical matter density, which barely expands forever; and a low-density $\Omega = 0.1$ universe, which easily expands forever. (Credit: J. Richard Gott and M. J. Rees, *Astronomy & Astrophysics*, 45: 365, 1975)

How do galaxies form? Basically by the same mechanism. The distribution of mass inside the cluster initially is not perfectly uniform. The little spherical region that will form the galaxy will have an *even higher* value of $\delta\rho/\rho$ than the cluster and will have a collapse time that is shorter than that of the cluster. Galaxies would form first, followed by clusters. Galaxies might have typical collapse times of about a billion years, whereas clusters of galaxies would have collapse times of several billion years.

Gunn and I were interested in what happens after the cluster formed. Consider a suburban galaxy, that is, one forming outside the spherical cluster, but close to it. The excess density inside the cluster proper also creates extra mass, pulling the suburban galaxy toward the center. The suburban galaxy will be decelerated by the gravity of the cluster but not by as much deceleration as experienced within the sphere itself. The cluster reaches its point of maximum expansion at a radius R_{max}. At that time the suburban galaxy is still moving outward, but later it will also stop (at a radius larger than R_{max}), turn around, and fall into the center of the cluster (after the time T_c). It will slosh through the center and find an equilibrium orbit, but one larger in size than the orbits of the galaxies in the original cluster. This infall will add mass to the cluster as the cluster acts like a gravitational vacuum cleaner, pulling in stuff from outside. We found that the mass of the cluster after the time T_c grew approximately like the ⅔ power of the time. This infall material created an extended envelope for the cluster that was less tightly bound and extended to larger distances. The density in the envelope fell off approximately like $1/r^{2.25}$. This was similar to the envelopes seen in great clusters.

Gunn and I applied our results to the Coma Cluster of galaxies. Zwicky had already estimated its total mass at about 10^{15} solar masses. Galaxies in this cluster were orbiting with typical velocities of about 1,700 kilometers/second. It had an envelope typical of what we might have expected from infall. In the center, the core of the cluster was dominated by two giant elliptical galaxies. Gunn and I were able to estimate their masses from their observed separation and their relative velocities. They weighed about 10^{13} solar masses, an order of magnitude more massive than our own galaxy. We then considered what might happen to the infall material. At that time people wondered if much of the mass of the universe could be hidden in the form of intergalactic gas. If that were the case, gas in the region beyond the cluster proper would infall as well as galaxies. When this gas slammed into the center of the cluster, it could not just pass through. The intergalactic gas is spread out, and its particles would tend to collide with other gas particles headed into the cluster from the opposite direction. This would heat the gas until the gas particles were also moving about the cluster with velocities of order 1,700 kilometers/second, reaching a temperature of about 10 million Kelvin, hot enough to glow in the X-rays. Any gas present in the cluster or in the surrounding region that fell in should be heated so that it would glow in the X-rays. When we started our research, no X-rays had been detected from clusters of galaxies. We thus had a prediction that could be tested. But before we could publish, such X-ray emissions were actually detected from clusters of galaxies. Given that, we had an explanation for this discovery already in hand, and we were eager to publish without delay.

A normal spiral galaxy like our own forms when a dense core collapses. The density in this core is relatively high when it reaches R_{max} and begins to recollapse. The star-formation rate typically depends on the square of the gas density, while the collapse rate depends on the square root of the density at turnaround at R_{max}. Thus, in this high-density central region, stars can form by the time the central region completes its collapse. The stars, being tiny, miss each other as they pass through the center and relax to form an elliptical *bulge*, where they orbit but are not drawn further in. Gas outside this region falls in later; it is at a lower density as it turns around and falls in and does not turn into stars before its collapse is completed. The gas heats up and then rapidly cools off, dissipating energy in the form of radiation. Because the gas cannot get rid of its rotational angular momentum, a rotating disk of gas is left. This gas disk can later slowly turn itself into stars. This produces a stellar disk (surrounding the central elliptical bulge), like the one in our own galaxy, which also contains some leftover gas, still forming new stars today.

In our own galaxy we see star formation still occurring. New high-mass stars are forming today—bright blue stars that light up the galaxy. As gas in the disk circles the center, gravitational traffic jams occur, creating spiral density waves. The gas is compressed as it enters the spiral density wave, and bright, short-lived, massive blue stars are formed. Being very bright, they use up energy at a prodigious rate and deplete their hydrogen fuel rapidly, dying quickly. The bright blue spiral arms are outlined by these short-lived stars. Lower-mass, lower-luminosity stars like the Sun and stars of even lower mass last longer and stay in the stellar disk. Thus, the disks of spiral galaxies are typically blue in color, dominated by the light of the bright blue stars. The bulge in the center of a spiral galaxy is redder. It is filled only with old low-mass stars, called *Population II* stars, as opposed to the *Population I* stars that are seen in the disk.

In the center of the Coma Cluster we see only elliptical galaxies and S0 galaxies—there are no spiral galaxies at all. S0 galaxies have elliptical bulges at their centers and disks of stars, but they have no gas currently forming stars. Gunn and I proposed that these galaxies used to be spirals that had fallen into the cluster and had had their gas stripped out as they plowed through the hot cluster gas. Their bright blue stars simply burnt out, and with the gas gone, no new stars could form to replace them. The hot gas in the cluster strips the gas from any spiral that falls in, turning it into an S0 galaxy. The burning out of the remaining blue stars occurs so fast (millions of years instead of billions) that it is hard to catch a dying spiral in the short interval before it turns into an S0. If two galaxies collided within the cluster, this could create a gravitational pileup, a jumble of stars that would later settle down to form an elliptical galaxy.

Jim Gunn and I emphasized that in the case of great clusters, the hot cluster gas was not going to cool in the age of the universe up to now, and so it would be left there within the cluster as hot gas, able to strip any spirals falling in. The cluster would gravitationally collect material from the neighborhood *outside* the cluster, and any intergalactic gas out there would be heated and show up in the X-rays. Intergalactic gas in the universe generally might be cool and invisible to us, but if it were out there, it would fall into clusters and heat up so we could see it in X-rays. This could give us a way of detecting and taking a census of intergalactic gas. Jim and I concluded that, given the X-ray observations, the universe did not have enough intergalactic gas ever to halt and reverse its expansion. The universe would expand forever. We called our paper, "On the infall of matter into clusters of galaxies and some effects on their evolution" (Gunn and Gott 1972). It is my most cited and Gunn's fourth-most cited paper, the first three being associated with data from the Sloan Digital Sky Survey (discussed in Chapter 9). It was a great privilege for me to work with Jim Gunn on this paper and others that we would do later. Jim went on to win numerous awards, including the MacArthur "genius" award, the Crafoord Prize, and the National Medal of Science. Jim may be said to be the person who really brought CCD digital cameras to astronomy; these cameras are about 100 times as sensitive as film and have revolutionized astronomy. He designed a CCD camera for the 200-inch telescope at Palomar Mountain, the Wide Field and Planetary camera on the Hubble Space Telescope, and the camera for the Sloan Digital Sky Survey.

Applications to Galaxy Formation

Jim Peebles at Princeton would subsequently show how galaxies got their angular momentum (rotation). As galaxies collapsed, they were not perfectly spherical. They would be subject to tidal forces from neighboring galaxies. If a galaxy had a barlike shape that was tipped at 45° relative to the line of sight to a neighboring galaxy, the end of the bar closest to the neighboring galaxy would be pulled more strongly toward it, while the end of the bar further away would be pulled less strongly. This would start the bar slowly spinning as it began to collapse. James Binney at Oxford then proved that elliptical galaxies forming from gravitational collapse would usually end up as triaxial ellipsoids—having elliptical shapes but with different diameters in three perpendicular directions.

In 1977, Jerry Ostriker (at Princeton) and Martin Rees (at Cambridge University in England) showed that for collapsing objects (like galaxies) smaller than 250,000 light-years across, the remaining gas would be sufficiently dense to cool and form a cold gas disk, while for collapsing objects larger than 250,000 light-years across (like the Coma Cluster), remaining gas would stay hot and not have time to cool off. This marked the difference between galaxies and clusters of galaxies. Gas falling into galaxies still in formation could cool and form a gas disk, which could then form stars and make a spiral galaxy. An especially dense galaxy could complete its star formation early and form an elliptical galaxy. Thus, galaxies form a distinct class of objects and then aggregate to form groups and clusters.

Gunn would apply our model of cluster expansion and collapse to the formation of our *Local Group* of galaxies, which is dominated by our own Milky Way and the Andromeda Galaxy. Originally, the Milky Way and Andromeda each formed from small fluctuations in the density distribution of the universe at recombination. Because of the excess density within each protogalaxy, each protogalaxy slowed its expansion relative to the rest of the universe and stopped expanding. The Milky Way and the Andromeda Galaxy then each collapsed and formed approximately a billion years after the Big Bang. Meanwhile the two galaxies continued expanding away from each other at a rate that was decelerated relative to the expansion of the universe as a whole. They eventually stopped moving away from each other and began to fall toward each other under their mutual gravitational attraction. The Andromeda Galaxy is now approaching us (one of the few galaxies to show a blueshift in its spectral lines, indicating a velocity toward us rather than away). It should collide with us about 4 billion years from now. After the stars of both galaxies have passed through each other and formed a jumble and the gas has collided, we should settle down ultimately to make one galaxy. Thus, as a cluster, our Local Group has a collapse time T_c, which is about 18 billion years—longer than the current age of the universe (13.8 billion years). Although the Local Group has not yet completed its collapse, it is still an appreciable density enhancement relative to the universe as a whole. Many loose clusters are in a similar state: loose associations of galaxies that are gravitationally bound but have not yet completed their collapse and have not reached a centrally condensed, relaxed state. These loose clusters do not have hot gas (glowing in X-rays) and are rich in spiral galaxies.

Jim Peebles believed that galaxies and clusters were formed "from the ground up"—smaller structures forming first. He knew that *globular clusters* (like those Shapley studied in the halo of our galaxy) had some of the oldest stars known, and he thought these globular clusters might be the first objects in the universe to collapse and form stars. He calculated that the first mass scale to be unstable to gravitational collapse after recombination would be the mass scale associated with globular clusters, about a million solar masses. These globular clusters would then cluster themselves to form the building blocks of galaxies. The galaxies would then cluster to form clusters of galaxies. Clusters of galaxies would draw together against the general expansion to form superclusters.

Bill Press and Paul Schechter at Caltech investigated this scenario via computer simulation. They started with seed masses of 3×10^7 solar masses distributed at random locations at recombination. These were just a bit larger than globular clusters. This random (*Poisson*) distribution of masses leads to random statistical fluctuations. For example, consider a region that should contain *on average* 100 masses. If you examine many such regions, you find that they typically contain 100 ± 10 masses.

(We call the 10 masses the *standard deviation* from the mean.) Sometimes they will contain as few as 90 masses and sometimes they will contain as many as 110 masses. When one contains 110 masses, it will constitute a fractional density enhancement of 10%, and will, according to the Gunn and Gott paper, collapse within a certain collapse time. If it has 90 masses, it will constitute an underdensity and will thin out. Eventually the collapsing centers will draw in this extra material by infall and add it to their own mass.

In a simulation starting with a random distribution of masses, a spherical region of sufficient size to contain *on average N* masses, will have typical fluctuations in this count of order $\sqrt{N}$. That gives typical *fractional* density excesses (or decrements) of order $(\sqrt{N})/N$. If a galaxy is defined as any object that collapses within 10^9 years, then (according to the Gunn and Gott formulas) this must start with the mass fluctuating upward by about 1% at recombination. Such fluctuations occur regularly in regions large enough to contain 10,000 of the seed masses on average. That's because $\sqrt{10,000} = 100$, and we expect these regions to typically contain $10,000 \pm 100$ masses, giving fractional density fluctuations of $\pm 100/10,000$, or $\pm 1\%$. Thus, we would expect typical galaxy masses to be of order 10,000 times the original seed mass (i.e., $10,000 \times 3 \times 10^7$ solar masses), or 3×10^{11} solar masses, which is about right—similar to the Milky Way (which is about 10^{12} solar masses).

Bill Press and Paul Schechter studied the distribution of masses for the collapsed galaxies that would result from such a simulation. In essence, they went searching in the initial conditions at recombination for the largest isolated spherical regions that had an excess density of 1% over the average. These would collapse in a billion years to form galaxies. They could figure out the distribution of galaxy masses that would result and write a formula for it. Schechter was able to fit the observed luminosity distribution of galaxies in large clusters well with a formula of this form. The Press and Schechter formula showed that random fluctuations in *mass* density could potentially explain the distribution of *luminosities* of galaxies that were observed, provided that the mass-to-light ratios of different galaxies were roughly equivalent, which seemed plausible. These mass perturbations were *isothermal* fluctuations—the temperature and density of the microwave background photons remained constant throughout, while the density of matter varied from place to place. It was a random distribution with fluctuations on a mass scale M at recombination of $\delta\rho/\rho = \pm 1\% \, (M^*/M)^{1/2}$, where M^* is the mass of a typical bright galaxy. On a mass scale of $16M^*$, the fluctuations would typically be about ±0.25%, for example. If these happened to be positive in sign (i.e., above average in density by 0.25%), they would be able to collapse in 8 billion years and form a cluster of 16 bright galaxies.

Hierarchical Clustering

Jim Peebles predicted how this clustering should go. A galaxy would form, and then it would tend to bond gravitationally with the galaxy nearest to it in the initial conditions. These would form a density enhancement, which would collapse to form a binary galaxy. The binary galaxy would bond with another binary galaxy to form a quadruple. Two quadruples would be pulled together to form an octuple, and then two of the octuples would form a cluster of 16 galaxies with a collapse time of about 8 billion years, according to Press and Schechter's simulation. There would be a hierarchy of clustering. Peebles and his graduate student Ray Soniera produced a hierarchical clustering model in just this way. They found that a galaxy today was likely to be in a tight binary, inside a quadruple galaxy system, inside an octuple—that is, hierarchically clustered. They placed the galaxies at random angles but at the proper distances to form the correctly bound hierarchy. If galaxies are clustered today, they should have more nearby neighbors than would be expected if galaxies were laid down at random today.

Peebles developed a very useful quantitative measure of this excess "neighborliness" among galaxies. If you sat on a random galaxy, the *covariance function* was defined as the average excess probability of finding a galaxy in a narrow shell of radius r away from you above and beyond the average density of galaxies in the universe. The covariance function was thus an elegant measure of the galaxy clustering. If the galaxies were initially in a Poisson distribution (distributed at random), as those point masses were in the initial conditions of the Press and Schechter model, the covariance function would be zero at all radii—no clustering on average initially. But in Poisson initial conditions, there are nevertheless random regions of excess density and underdensity with $\delta\rho/\rho = \pm 1\%$ $(M^*/M)^{1/2}$. The galaxies in the random density *enhancements* would be drawn together by gravity and eventually collapse to form clusters. Peebles then did a little fancy algebra,[2] and he concluded that if you are sitting on a typical galaxy living in this hierarchy of binaries-within-groups-within-clusters and you look out, the number density of galaxies you see around you should fall off like $r^{-1.8}$. Since the clusters are large-density enhancements, the excess density is nearly equal to the total density (the mean density of the universe being small by comparison), so the covariance function should be proportional to $r^{-1.8}$.

Peebles then went ahead and measured the covariance function of galaxies in the real sky (using the Zwicky and Shane-Wirtanen catalogs of galaxies). By mathematical inference he could deduce from the projected positions of the galaxies on the sky their covariance function in 3D. This is possible because the projected separation between two galaxies on the sky is statistically related to their 3D separation. Averaging over all galaxies, Peebles found that the covariance function—denoted as $\xi(r)$—was $\xi(r) \approx (r/24 \text{ million light-years})^{-1.77}$. Rounding to two significant figures, it was proportional to $r^{-1.8}$, matching the theoretical calculation of what a hierarchy of clusters formed out of Poisson initial conditions should produce! This agreement reinforced the idea that the (actual) initial conditions were a random (Poisson) distribution of seed masses.

The covariance function was equal to 1 at a radius of 24 million light-years. This meant that if you sat on any galaxy and looked out to a distance of 24 million light-years away from you in any direction, you would find *on average* about 1 + 1, or twice the average number density of galaxies found in the universe. You would find in a thin radial shell at that radius about twice as many galaxies as you expected. (By the way, the mean separation between galaxies in the universe is also about 24 million light-years—an interesting coincidence.) One has to add up the result over many galaxies and take the average. Peebles' theoretical calculation was based on an $\Omega_m = 1$ Friedmann universe, where the matter density was equal to the critical density, one that had barely enough kinetic energy in its expansion to overcome the gravitational attraction of the matter and continue to expand forever. Peebles thought the fact that this model produced a covariance function matching what was observed supported the idea of a universe with a critical mass density—pretty simple and elegant, all the way around.

Jim Peebles came to Caltech to give a colloquium. I was there as a postdoc. Peebles had done the world's largest *N-body* computer simulation. He had placed 1,000 point masses ($N = 1{,}000$), each representing a galaxy, initially sprinkled at random (a Poisson distribution) inside a spherical volume. Then he started the sphere off with a Hubble expansion, adjusted so that the universe met the conditions of the critical density case. The radius of the sphere would then grow as the ⅔ power of the time. This means that the sphere grew in size by a factor of 4 as the time since the Big Bang increased by a factor of 8 (because $4 = 8^{2/3}$). The expansion gradually slowed down because of the mutual gravitational attraction of the galaxies. As the sphere expanded, the galaxies began to cluster just as Peebles had claimed. He made a movie whose scale expanded with time in synch with the expansion of the sphere. Thus, on the screen, the sphere of galaxies remained the same size. But you could watch with time as the galaxies went from their initial Poisson random distribution to the clustered distribution we see today. It was like watching the history of the universe! First, nearby galaxies fell together and made tight binaries. Soon, small groups could be seen forming, and then groups fell together to form clusters. All the while, Jim was standing in front of the screen, dramatically pointing to the places where he knew a cluster was about to form, giving a play-by-play like a sportscaster. Then, Richard Feynman shouted out, "Get away from the screen, let us see it!"—gesturing for Peebles to move aside so he could get a better view. He was clearly enjoying the show and was smart enough to figure it out for himself without commentary! By the end, there were even great clusters like the Coma cluster. It was quite a success.

Two astronomers at the Lick Observatory, Donald Shane and Carl Wirtanen, had carefully counted galaxies on photographic plates of the sky for years. (I've already mentioned how Peebles made use of their catalog to calculate the covariance function in the universe.) The Lick survey covered about one quarter of the entire sky centered on the North Galactic Pole. They had counts of galaxies in small bins covering the whole survey region. In all, the two had laboriously counted a million galaxies. They would patiently count galaxies on these glass plates, all day long, day after day. In the end, they were so exhausted that, except for the counts they had obtained, they had little time to study their own catalog! Jim Peebles and his colleagues entered the data into a computer so that a digital image could be constructed. On the largest scales the distribution was uniform (as Hubble had noted), with approximately as many galaxies on the left side of the picture as on the right side. But on smaller scales the clumping of galaxies was obvious. Most prominent in the picture, near the center, was the dense kernel of the Coma Cluster. Peebles could now compare this picture with the hierarchical clustering model he and Soniera had produced, which also had a million galaxies. The agreement was quite impressive. The hierarchical model reigned supreme. The universe seemed to be described best as clusters within clusters within clusters—or "meatballs" (as we would nickname them) within meatballs within meatballs—all floating in a low-density sea. Clusters formed from positive-density fluctuations in the initial conditions, and they grew by drawing in additional material from adjoining low-density regions. The meatballs got more and more dense relative to the background density. The message was: "Keep your eye on the clusters—that's where the action is." This was the view of the American school of cosmology. Before long, all the American astronomical community had seen Peebles' movie and his results on the covariance function, and they were impressed.

The hierarchical model would ultimately be topped off by the 1983 paper of Neta Bahcall and Ray Soniera, who showed that great clusters were themselves clustered; these great clusters had a covariance function with other great clusters that was a power law with the same slope as the galaxy covariance function: $\xi(r) = (r/120 \text{ million light-years})^{-1.8}$. This result is exactly what one would have expected from the hierarchical model of Peebles and Soniera. There were clusters of clusters—*superclusters* of galaxies. As early as 1953, Gerard de Vaucouleurs had noticed that the Virgo cluster of galaxies had extra galaxies in its neighborhood with the number density of galaxies falling off like $r^{-1.7}$ as a function of radius measured from the Virgo Cluster center. The Virgo Cluster was accompanied by surrounding galaxies. De Vaucouleurs ultimately called this our Local Supercluster. Today it is known as the

Virgo Supercluster. Its diameter is about 100 million light-years, and our own Local Group of galaxies is in its outskirts. Our address in space is: Earth, Solar System, Milky Way, Local Group, Virgo Supercluster: clusters within superclusters, meatballs within meatballs.

A Year at Trinity College, Cambridge University

After Caltech, I went to Cambridge University and worked with Martin Rees, who was head of the Institute of Astronomy at that time. He found me a place at Trinity College, one of the many colleges that make up Cambridge University. Trinity College was Isaac Newton's old college. It was quite a privilege to be there. As a visiting Fellow Commoner, I ate at the high table every night with many distinguished scientists. I got to know Alan Baker, a number theorist who had won the Fields Medal, Brian Josephson, who invented the Josephson junction, and Douglas Heggie, a Scottish astronomer with an interest in Paleolithic astronomy. The senior fellows sat at the head of the table and escorted us upstairs to port after dinner every night. Some of the fellows were in their nineties but still had plenty of energy to share reminiscences. The most senior fellow was Lord Adrian, who had discovered the reflex reaction that gets you to lift your finger off a hot stove before the nerve signal gets to your brain. Next to him sat Littlewood, the famous mathematician, who would tell stories of his legendary colleagues Ramanujan and Hardy. Next to them was Mr. Nicholas, the healthiest 87-year-old I had ever seen. He had been a geologist and had a ruddy complexion and perfectly white hair; except for the white hair, he would have looked 65. As a junior fellow, he had lived in the college until he married at age 30. He and his wife celebrated their fiftieth anniversary just before she died, when he was 80. He showed me her picture. He had moved back into the college after she was gone and had been living there again for 7 years when I met him. Once I asked him if he had seen Halley's Comet when it appeared in 1910. Yes, he had. But as a young man he also remembered talking to an old fellow of the college who had seen it on its previous visit, in 1835! Such were the conversations one could have at Trinity College. Ultimately, Mr. Nicholas lived to be 101, still entertaining the fellows with stories on his 100th birthday.

At times during my year at Trinity, I felt as though I was living in a starship that had been traveling toward its goal for centuries. I could visit Newton's rooms, where he had written the *Principia*, outlining his theory of gravity. Newton's own copy of the book, with his marginal notes for the second edition, was on display in the college library. In the antechapel stood a statue of Newton, honored by Wordsworth with these lines from his poem *The Prelude* (III.58–63):

And from my pillow, looking forth by light
Of moon or favouring stars, I could behold
The antechapel where the statue stood
Of Newton with his prism and silent face,
The marble index of a mind for ever
Voyaging through strange seas of Thought, alone.

The Great Court of the college looked much as it had in Newton's day. In the center was the fountain where, legend has it, young Lord Byron used to tether his pet bear!

Martin Rees lived down the road in Kings College (renowned for its magnificent chapel and choir), in rooms formerly occupied by the world-famous economist John Maynard Keynes. In 1975 Martin was a young, but already distinguished, professor. (He would later become Astronomer Royal of England and Master of Trinity College, be knighted, and become Lord Rees and then Baron Rees of Ludlow.) Our 1975 research concerned large-scale structure. We were trying to find out where those very early isothermal fluctuations, needed by Peebles and by Press and Schechter, really came from. Martin was familiar with the work of Yakov Zeldovich and the Russian school of cosmology. Zeldovich was working on large-scale structure from a completely different angle. He realized that in the standard Big Bang model, we continue to see larger and larger regions of the universe as time goes on. When the universe is 1 second old, we can see out to a distance of 1 light-second—the distance light can travel in 1 second. When the universe is 1 year old, we can see out 1 light-year. Now, when the universe is 13.8 billion years old, we can see out to a distance of 13.8 billion light-years. The universe we see on the largest scales is nearly, but not perfectly, uniform. To first order, the universe on large scales is approximately uniform with similar counts of galaxies in different directions, but superimposed on this uniformity are small fluctuations. Without fluctuations, the universe would remain perfectly uniform and no galaxies or stars would ever form. We would not be here. We needed a uniform universe plus small fluctuations.

Zeldovich postulated that the ratio $\delta\rho/\rho$ (density fluctuations/average density) on the largest scales visible should remain constant as the universe expanded, as we saw larger and larger regions with time—that is, as light had more time to travel to us from distant realms. It was a bold hypothesis. It was simple. Martin and I applied this to isothermal fluctuations (fluctuations in matter density alone while the radiation temperature and density remained constant—like Press and Schechter had postulated). We realized that as the universe expands, the matter density goes down as the cube of the expansion factor. If the universe expands by a factor of 2, the density of matter goes down by a factor of 8 as the matter particles fan out into a volume $2 \times 2 \times 2 = 8$ times larger. The energy density of the cosmic microwave background radiation goes down like the fourth power of the expansion factor. As the universe expands by a factor of 2, the number density of photons in the microwave background decreases by a factor of 8. But the wavelengths of these photons are also stretched by a factor of 2 because of the stretching of space. The energy of a photon is, by Einstein's formula, proportional to 1 over its wavelength. Therefore, when the universe expands by a factor of 2, the energy of each photon drops by a factor of 2 as well. If the number density of photons drops by a factor of 8 and the energy of each photon also drops by a factor of 2, then the overall energy density of the cosmic microwave background radiation must drop by a factor of 16 (or 2^4). This means that the energy density in the thermal radiation falls faster than the matter density as the universe expands.

Conversely, if we trace backward in time to the moment of the Big Bang, we will find the radiation density increasing faster than the matter density. As we get close enough to the Big Bang, the radiation must become dominant. George Gamow understood this. When the mass scales associated with galaxies and clusters of galaxies come into view after the Big Bang, we are still at early times when radiation is dominant. The mass in matter we can see goes up as a function of time as we get further and further away from the Big Bang and light has had more time to travel. The ratio of matter density to radiation density in the universe is also increasing with time. If Zeldovich was right, Martin and I reasoned, total $\delta\rho/\rho$ on the largest visible scales would stay constant with time. If the perturbations were isothermal, the radiation would be the same temperature everywhere at a given epoch and would have no intrinsic fluctuations itself. The $\delta\rho/\rho$ in total density would have to be produced by fluctuations in the matter alone. At earlier times when the matter was a smaller part of the total density, the fluctuations in the matter would have to be larger in order to keep the total density fluctuations at the same Zeldovich amplitude. At earlier times, one could see out only a short distance because of the finite velocity of light, so one could see only a small amount (mass) of matter. This would mean that the fluctuations in the matter on small scales (visible at earlier times) had to be larger than the fluctuations in the matter on large mass scales (which became visible later). Martin and I calculated that $(\delta\rho/\rho)_{\text{matter}}$ should be proportional to $M^{-1/3}$, where M is the mass scale of the fluctuation. This was close to the $(\delta\rho/\rho)_{\text{matter}}$ proportional to $M^{-1/2}$ in the initial conditions that Peebles got for his Poisson fluctuations. These matter fluctuations would remain frozen with the cosmic microwave background as the universe expanded until recombination. Matter is coupled tightly to the photons because the matter is ionized, so there are negatively charged electrons plus protons, primarily. By the time recombination occurs, the temperature of the radiation has cooled enough for the electrons and protons to combine to make neutral hydrogen gas, which is free to move relative to the radiation. Gravitational clustering takes over, just as in Peebles's computer simulation.

The small difference between the $M^{-1/3}$ relation that Rees and I found and the $M^{-1/2}$ relation Peebles used was actually helpful. The catalog of groups of galaxies compiled by Ed Turner and myself at Caltech had detected many binary galaxies and small groups while also finding the famous Virgo cluster of galaxies. The observed distribution of groups and clusters of galaxies was broader in mass than the Press and Schechter formulas would have suggested if $(\delta\rho/\rho)_{\text{matter}}$ in the initial conditions was proportional to $M^{-1/2}$, but the data were fit well by the $M^{-1/3}$ relation. Also, Peebles's calculations for the covariance function had been calculated on the basis of a cosmology having critical density. But observations suggested that the universe actually had a matter density that was less than the critical density. Ostriker, Peebles, and Yahil (1974), for example, had estimated it as one-fifth the critical density. In such a low-density universe, growth of structure at early epochs was similar to that in a critical-density model, but structure grew less at late epochs when the matter density began to depart significantly from the critical density as the matter thinned out. Thus, in such low-density universes, there was less growth of structure at large scales, relative to small scales. If the initial conditions were Poisson with the $M^{-1/2}$ distribution, this would make the covariance function too steep in a low-density universe. But if one started with an $M^{-1/3}$ distribution, this would compensate for less growth at large scales in a low-density universe. Martin and I concluded that the covariance function in a low-density universe would be okay if one started with fluctuations at recombination in the matter, $(\delta\rho/\rho)_{\text{matter}}$, proportional to $M^{-1/3}$ (see Figure 3.2).

The other kind of density fluctuations people often discussed were *adiabatic fluctuations*—fluctuations in the density of the thermal radiation in the hot Big Bang accompanied by equal fluctuations in the matter density. A region that had more photons also had more protons, neutrons, and electrons as well. Thermal radiation has an energy density, and according to Einstein's equation $E = mc^2$, this corresponds to a certain mass density. Fluctuations in the temperature of the radiation cause fluctuations in the energy density of the radiation. Unfortunately, these adiabatic fluctuations got erased at mass scales smaller than the Silk Mass (10^{12} to 10^{14} solar masses) as we shall discuss in the next chapter. In 1974, Doroshkevich, Sunyaev, and Zeldovich had shown that adiabatic fluctuations (obeying Zeldovich's hypothesis of constant amplitude on the largest scales becoming visible as a function of time) would also produce matter fluctuations proportional to $M^{-1/3}$ on mass scales *larger* than the Silk Mass (see Figure 3.2). Thus, if one had only adiabatic fluctuations in the matter plus radiation occurring equally and in synch, then on very large scales the fluctuations should also scale as $M^{-1/3}$. General density fluctuations (in the matter and in the radiation) could always be decomposed into isothermal and adiabatic components, so it was promising that both components suggested matter fluctuations proportional to $M^{-1/3}$ on the largest mass scales. But Martin Rees and I wanted to focus on isothermal fluctuations because these could extend down to small scales and allow us to make galaxies first, as Peebles had argued. We wanted to make galaxies first and then have them cluster.

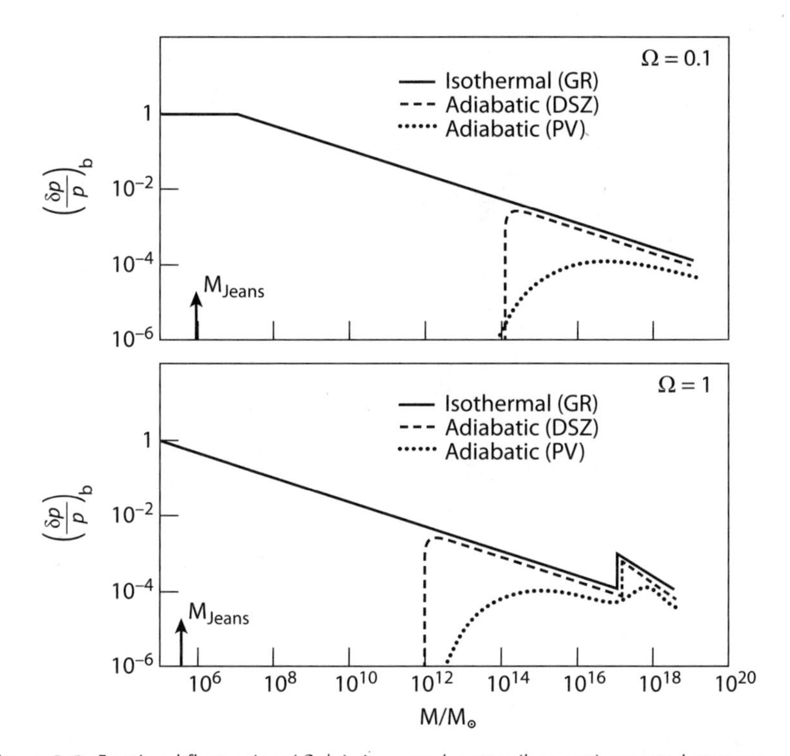

Figure 3.2. Fractional fluctuations$(\delta\rho/\rho)_b$ in normal matter (baryons) expected at recombination from different models (vertical axis) as a function of mass scale in solar masses (horizontal axis) in low-density (Ω = 0.1, top panel) and high-density (Ω = 1, bottom panel) universes. These calculations (circa 1979) follow Zeldovich's hypothesis that fluctuations are of constant amplitude ($\delta\rho/\rho = 10^{-4}$) as they first become visible. GR—(Gott–Rees) isothermal fluctuations; DSZ—(Doroshkevich–Sunyaev–Zeldovich) adiabatic fluctuations with instantaneous recombination; and PV—(Press and Vishniac) adiabatic fluctuations with a realistic finite timescale for recombination. M_{Jeans} is the smallest mass unstable to collapse after recombination. Adiabatic fluctuations on scales smaller than the Silk mass (from 10^{12} to 10^{14} solar masses) are damped out (erased), and can't be used to make galaxies (10^{12} solar masses) directly. (Credit: J. Richard Gott, Lecture Notes of Les Houches summer school, 1979)

I was able to check some of these results by doing large *N*-body computer simulations at Cambridge with Sverre Aarseth and Ed Turner. Besides being one of the world's top experts on large *N*-body simulations, Aarseth was an accomplished mountaineer, having summited a number of high Himalayan peaks, and he was one of the 10 best postal chess players in the world—quite an interesting fellow! Aarseth had developed a very sophisticated *N*-body code. Usually, when a tight binary formed,

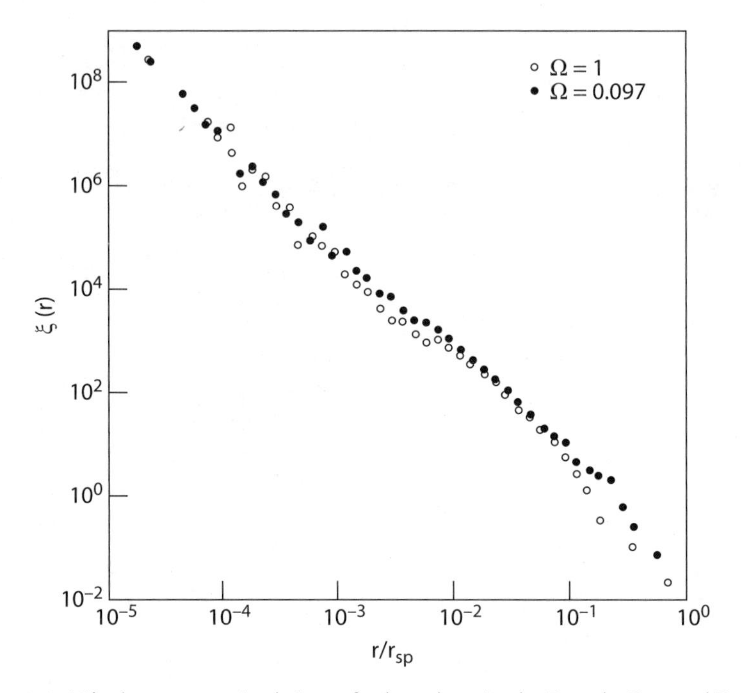

Figure 3.3. *N*-body computer simulations of galaxy clustering by Aarseth, Gott, and Turner produced these covariance functions of galaxies as a function of radius ($\xi(r)$). The radius of the simulation volume is r_{sp}. Results at the present epoch from two simulations are shown: a high-density ($\Omega = 1$, open circles) universe with initial conditions having $(\delta\rho/\rho)_{matter}$ proportional to $M^{-1/2}$, as Peebles proposed; and a low-density ($\Omega = 0.097$, filled circles) universe with initial conditions having $(\delta\rho/\rho)_{matter}$ proportional to $M^{-1/3}$, as Martin Rees and I proposed. The two results are essentially identical, and close to the observations as measured by Peebles. (Credit: S. J. Aarseth, E. L. Turner, and J. Richard Gott, *Astrophysical Journal*, 228: 664, 1979)

one had to slow down the whole calculation as one took tiny time steps to follow this rapidly orbiting pair using Newton's laws. Aarseth simply computed the orbital elements for the Keplerian elliptical orbit that the binary would have, allowing him to return to it many orbits later when the big simulation had taken another normal time step. We were thus able to follow the evolution of 4,000 particles in an expanding cosmology. This was 4 times larger than the simulation Peebles had shown at Caltech. We found that a low-density ($\Omega_m = 0.097$) universe with initial $(\delta\rho/\rho)_{matter}$ proportional to $M^{-1/3}$ was able to produce a covariance function proportional to $r^{-1.9}$, close to the observed $r^{-1.8}$ relation. A high-density ($\Omega_m = 1$) universe with initial $(\delta\rho/\rho)_{matter}$ proportional to $M^{-1/2}$ gave nearly identical results. See Figure 3.3.

These were the models and lines of evidence the Americans were considering. We were following the lead established by Jim Peebles: a "meatball" universe—isolated clusters forming and growing by gravitational attraction in a low-density soup: meatball soup. Later, Jim Peebles, Jim Gunn, and Martin Rees would share the Crafoord Prize awarded by the Swedish Academy for their contributions to understanding large-scale structure. But in the Soviet Union, Zeldovich and his colleagues were cooking up a different story.

Chapter 4

The Great Void in Boötes—A Swiss Cheese Universe

Yakov Zeldovich was the dean of Soviet astrophysics. He was also a father of the Soviet hydrogen bomb. After World War II, the Soviets were quick to develop an atomic bomb to match the Americans, just as Leo Szilard, the inventor of the nuclear chain reaction, had feared. Szilard was the one who had convinced Einstein to write the fateful letter to Roosevelt to start the Manhattan Project. Szilard knew that the physics for making such a bomb was within reach and Hitler might well get one first. After all, Otto Hahn in Germany had split the atom. The Manhattan Project was begun in the United States, and its participants worked at full speed to have an atomic bomb available as soon as possible. But Hitler was defeated before the U.S. atomic bomb was ready. Szilard was not in favor of using it against the Japanese, who did not have the bomb, he thought, and would be defeated anyway. He argued that even testing the bomb was dangerous, as it would cause the Soviets to develop one as well and initiate an arms race. He feared this would ultimately lead to a third world war, this time fought with atomic weapons. Szilard tried to deliver a petition to Truman from noted scientists arguing that the atomic bomb should not be tested and not be used against Japan. He gave the petition to a southern senator who promised to get it to Truman, but Truman never saw it. After the war, the United States continued an assembly line producing additional atomic bombs.

It was soon realized that an atomic bomb could be used as a trigger for an even more devastating hydrogen bomb. The atomic bomb worked by nuclear fission, splitting uranium, or plutonium nuclei. The hydrogen bomb worked by fusing hydrogen nuclei together to form helium. Both reactions released energy by Einstein's famous equation, $E = mc^2$. But the release of energy in fusing hydrogen into helium was much greater. This fusion reaction is what powers the Sun, and such reactions also occurred in the early universe.

I had first learned in detail of Zeldovich's work on this topic when I was in a "shotgun" journal club run by Jim Gunn at Caltech. (While I was a graduate student, Gunn moved from Princeton to Caltech; I was very happy to go there to have a further opportunity to work with him again as a postdoc.) The shotgun nature of the journal club was such that everyone read that week's paper and prepared a talk on it. Then a name was drawn from a hat and that person had to give the talk to the others, who were prepared to ask questions. It was nerve-racking. My name was drawn to give the talk on Zeldovich's paper on detonation waves in supernovae. But we all realized that the same physics could apply in theory to detonation of a hydrogen bomb, where an atomic bomb at the center formed the trigger. It would heat the hydrogen at the center to a temperature hot enough for hydrogen to fuse. The hydrogen fusing would create a high-enough temperature for the hydrogen in the next shell outward to fuse as well. A conflagration blast wave would proceed outward like a forest fire, burning all the hydrogen. Add more and more hydrogen to your bomb, and you could make it as big as you wanted.

The United States would eventually test hydrogen bombs with an explosive power equal to 15 megatons of TNT. But the Soviets would build and test the largest hydrogen bomb, with a power of 57 megatons of TNT. This was 2,850 times as powerful as the 20-kiloton-TNT-equivalent atomic bomb dropped on Hiroshima. Zeldovich, with his intimate knowledge of the Soviet hydrogen bomb, was not allowed to travel outside the Soviet Bloc. That is one reason his work in astrophysics was not as widely known outside the USSR as it should have been at that time. Sometimes, to convince people of your ideas, you need to pour them directly into their ears by giving talks and attending conferences. But Martin Rees did travel widely, and he was impressed by Zeldovich's work on large-scale structure. I became acquainted with it during my time in Cambridge.

Zeldovich believed only in adiabatic fluctuations. They made more sense physically. Adiabatic fluctuations were just variations in the temperature of the thermal radiation in the universe from place to place. If there were more photons in one region, there should be more protons, neutrons, and electrons in that region as well. They just followed the photons. That made sense if one started with pure thermal radiation and the excess of matter over antimatter in the universe prior to 1 second after the Big Bang was the result of the asymmetric decay of massive particles (favoring decay into matter slightly over decay into antimatter) in the very early universe. Hotter regions would have more photons and more of every type of particle as well. The American model as exemplified by Press and Schecter instead called for isothermal fluctuations, in which only the protons, neutrons, and electrons varied from place to place, while the photons remained mysteriously uniform. Zeldovich found this model to be ad hoc and relied on adiabatic fluctuations instead. Here's how it worked. Put photons into a box and, because they are moving at the speed of light, they will hit the sides of the box, creating a pressure equal to one-third the energy density of the photons. So, if you squeeze radiation, it rebounds and creates a sound wave. The electrons (and the protons they drag along) are coupled to the radiation and participate in the sound waves. These just oscillate, as sound waves do, rather than having a chance to grow in amplitude. Photons would be able, by slow diffusion, to random walk out of any small perturbations (what is called the Silk Mass of 10^{12} to 10^{14} solar masses) over time and damp these perturbations down in amplitude—dragging the nuclei and electrons along with them. This would erase fluctuations smaller than the Silk Mass. If neutrinos had mass, as Zeldovich hypothesized, they would participate in the adiabatic fluctuations. Being weakly interacting, these neutrinos would decouple from the radiation and be free to cluster due to gravity at earlier times than the matter, which was electromagnetically coupled to the radiation. This could be helpful in explaining galaxy clustering, but the velocities of the neutrinos would be large (they would be called "hot dark matter") and they could freely stream out of any galaxy-mass initial fluctuations. Thus, in Zeldovich's mind, fluctuations on galactic mass scales would be erased. If so, how do galaxies form? Zeldovich proposed that large-scale structure would have to form first, and galaxies later.

Gunn and I had examined what would happen to a spherical region with excess density. It would stop expanding and eventually recollapse to form a cluster at the center. But Zeldovich considered what would happen if the density fluctuation were not quite spherical in shape—suppose it was a slightly flattened sphere? As it expanded, it would be more decelerated along the short axis and would become more elliptical. Eventually it would even stop expanding along the shortest axis and start collapsing in that direction while still expanding in the other two directions. It would become a giant pancake. When its thickness collapsed to zero, and it indeed became flat as a pancake, its density would soar, and galaxy formation would be triggered. Any gas falling in, hitting the pancake from both sides, would be heated, but since it was at high density, it would be able to cool quickly. Galaxies could then form in the thin pancake, Zeldovich argued.

The Voronoi Honeycomb

The *Zeldovich pancakes* represented high-density surfaces formed by gravity, where gas from different regions had crashed together. These are called *caustic surfaces*. Breaking waves in the ocean form caustic surfaces, for example. What would the geometry of these pancakes look like in the universe as a whole? How would the pancakes connect up? Zeldovich, Jaan Einasto, and Sergei Shandarin investigated this. They found that the universe would come to resemble a giant honeycomb with empty cells surrounded by walls (pancakes). The high-density walls in the honeycomb would be all in one connected piece, whereas the voids would be isolated—separated from each other by cell walls. This is the opposite of the meatball topology (considered in Chapter 3), in which the high-density regions are all separate and the low-density regions are connected in one piece. A honeycomb, with empty cells, has the same topology as Swiss cheese (as our topology group noted in America). The voids are isolated, while the cheese itself (the high-density part) is all in one piece.

The random cells in this honeycomb would approximate a mathematical distribution called a *Voronoi tessellation*. Just as squares will tile the plane in an infinite checkerboard pattern, you can stack cubical

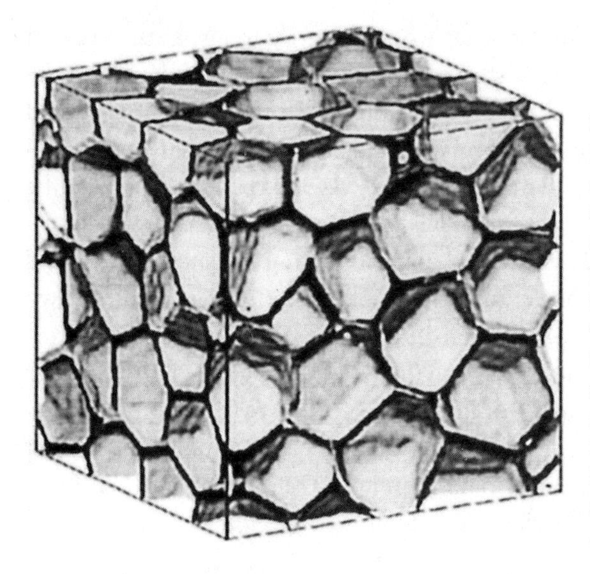

Figure 4.1. A Voronoi honeycomb. Its cells have random centers. The walls of the cells form the honeycomb and, in Zeldovich's picture, galaxies formed in the walls. This left giant voids in the galaxy distribution. (Credit: A. P. Roberts and E. J. Garboczi, *Acta Materialia*, Vol. 49 (2): 189, 2001, Figure 2, located at http://ciks.cbt.nist.gov/garbocz/closedcell/node5.html)

cardboard boxes to fill an infinite warehouse with no gaps. The boxes are empty, making cells of a cubical honeycomb. The cardboard material of the empty boxes stacked together forms one piece—a cardboard honeycomb. But this is a regular honeycomb. We expect the universe to have a randomness in the shapes of its cells. So here is how to make a random Voronoi honeycomb.

Sprinkle points randomly in three-dimensional space. These points will denote the "centers" of cells. Label these centers 1, 2, 3, 4, ... Cell 1 is the set of points in 3D space closer to "center" 1 than to any other center. Cell 2 is the set of points in 3D space closer to center 2 than to any other center, and so forth. If you want to know which cell you belong to, ask which cell center is closest to you. What do the boundaries of the cells look like? The set of points equidistant from two points, or centers, in 3D space is a plane that bisects and is perpendicular to the line connecting the two points. Thus the boundaries of the cells are segments of planes. These are the Zeldovich pancakes. The cells are irregular polyhedrons, bounded by flat planar faces. These planar faces meet at straight edges, which in turn meet at vertices. If you live on the boundary of a cell, you are equidistant between two cell centers and can't tell which cell you belong to. Figure 4.1 shows a Voronoi honeycomb.

The galaxies live on these planar faces. They form there when the pancakes collapse. At an edge, three or more planar faces in the honeycomb come together and galaxies can, after they are formed on the planar faces, fall together by gravity to form a filament of galaxies along the edge. In a map of the United States, typically three states come together randomly to form a vertex. (There is only one exception to this, Four Corners, where Arizona, New Mexico, Colorado, and Utah come together.) But usually, three states come together at a point—which explains why so many local TV newscasts are geared to a tristate area. Likewise, in a random network of polyhedral cells filling space, three cells will typically come together at an edge. That means three flat cell walls typically join at an edge. This gives more galaxies, lying in the vicinity of an edge, galaxies that can fall together by the action of gravity to make a dense filament or chain of galaxies along that edge. At a vertex, typically four cells will meet up, and six walls and four edges will meet there, producing even more gravitational attraction.

Zeldovich thought the vertices were where the great clusters like the Coma Cluster would form. So voids would be empty, surrounded by thin planes of galaxies; these would by outlined by heavier filaments of galaxies, with great clusters of galaxies at the vertices. Kepler would be happy that polyhedrons were important in understanding the universe after all! This is the Zeldovich universe: a honeycomb with cells—a Swiss cheese universe.

I first met Zeldovich when I went to Tallinn, Estonia, in the USSR for a meeting sponsored by the International Astronomical Union in 1977. He came up to me and said: "Gott, three—I know you. What is the three?" I explained that my full name is John Richard Gott, III; my father is John Richard Gott, Jr.; and my grandfather was John Richard Gott. So I'm the third. In my astronomy papers I went by J. Richard Gott, III, so that is what he remembered. As a book author, for simplicity I just go by J. Richard Gott, since that is how most people cite my name. Zeldovich went on to say: "You look like the young Landau!" I hadn't thought of it but it was true—I did rather look like Lev Landau, the great Russian physicist and 1962 Nobelist. Like him, I had (back then!) a shock of dark wavy hair parted to one side, and we had a similar facial shape and eyes. I was pleased to be compared to him, if only in appearance! At the conference I described work Ed Turner and I had done on the mass-to-light ratios of groups of galaxies. We used median mass-to-light ratios to eliminate spurious values resulting from contamination by background or foreground galaxies or unlucky cases where the velocities of the galaxies are pointed in special directions relative to our line of sight, which can throw the values off. Our results favored a low-matter density universe that would expand forever. Zeldovich complimented our work after the talk. "Yes, always use the median," he said. "In Russia the watches are not made very well, so when friends get together they compare the times on their watches. One says 5 minutes to 5, another says 5 o'clock, the other says 11 o'clock. Use the median!" I don't think anyone has ever summarized the advantage of using the median better.

I would later write a paper with my colleagues on median statistics, using them to reconfirm the acceleration of the universe just discovered by two competing groups. We showed that the supernovae they were seeing were indeed faint, not because one or two outliers were making the average fainter, but because most of them were surprisingly faint. That increased one's confidence in their result that the expansion of the universe was accelerating. In that same paper we applied median statistics to the Hubble constant. Hubble himself had found a value of 500 kilometers/second/megaparsec. Sandage found 55 ± 3 and de Vaucouleurs found 90 ± 10, while the Hubble Space Telescope team found a value near 74. Gravitational lensing measurements were typically in the 60s, and even lower values emerged from the Zeldovich-Sunyaev method of comparing X-ray luminosities of galaxy clusters with the diminution in the cosmic microwave background radiation seen behind them, which was caused by scattering of the microwave background photons by their hot gas. Different methods gave discordant estimates. We simply took the median and got 67 kilometers/second/megaparsec. That was in 2001. In 2013, the Planck satellite team took the best available data, including methods using fluctuation scales seen in the cosmic microwave background, and concluded that the Hubble constant was 67 ± 1 kilometers/second/megaparsec. (The latest results, as of 2015, are 67.8 ± 0.9 from the Planck satellite team and 67.3 ± 1.1 from the Sloan Digital Sky Survey.) We were 12 years ahead of time—thanks to encouragement from Zeldovich!

Observational Evidence

Some observational evidence indicated that the Zeldovich honeycomb model for the universe might be right. First, Gérard de Vaucouleurs had found that the Virgo Supercluster, of which our own local group and the Virgo Cluster were members, had a flat geometry. If Zeldovich pancakes were common, we should be sitting on one of them, assuming our galaxy is not special.

In a 1980 paper, Jaan Einasto, Mihkel Joeveer, and Enn Saar, using available redshift data, summarized the distribution of galaxies in space: "The densest condensations of matter, rich clusters . . . have been joined by slightly curved or straight chains of clusters of galaxies. . . . The planes joining neighboring chains are also populated by galaxies. The whole picture resembles cells. One cell wall with surrounding cluster chains can be taken for a supercluster. The space inside voids is void of clusters and almost void of galaxies. Arguments have been given suggesting that superclusters formed prior to the formation of galaxies or simultaneously with them." This completely supported the Zeldovich picture.

Perhaps most spectacularly, in 1981, a group of astronomers—Robert Kirshner, Augustus Oemler, Paul Schechter, and Stephen Shectman—found a million-cubic-megaparsec void in the constellation of Boötes. That's a cube 326 million light-years on a side. This region was virtually devoid of bright galaxies, and it was breathtakingly large. In three (very narrow) pencil-beam surveys about 35° apart in the sky, they found only 1 bright galaxy in the void, whereas they would have expected to see about 25 galaxies. These three narrow-beam surveys covered only three small regions in the sky, rather like taking three core samples in geology, looking for oil. But here they were looking for galaxies and not finding as many as expected.

The void existed between recession velocities of 12,000 and 18,000 kilometers/second. In front of the void, where galaxies are receding from us at about 9,000 kilometers/second, the galaxy distribution has a peak—many galaxies sitting on the front edge of the void at a distance of about 440 million light-years away from us. We determine that distance by using Hubble's law: $d = v/H_0$, where d is the distance of the galaxies in megaparsecs (3.26 million light-years each), v is the recession velocity of the galaxies, and H_0 = 67 kilometers/second/megaparsec. Thus d = (9,000/67) × 3.26 million light-years = 440 million light-years. Beyond the void there is a spike in galaxy counts at a recession velocity of 21,000 kilometers/second, or at a distance of 1,020 million light-years. Thus, there are many galaxies in a narrow band at the front edge of the void and at the back edge. This is just what the Zeldovich model would predict—pancakes of galaxies at the front and back of a nearly empty cell. Assuming their three "test wells" are typical, this void has about 1⁄25 the normal number density of galaxies, so a huge region that would normally contain 2,400 galaxies actually contains less than a hundred, as if it were a relatively empty honeycomb cell. This was a complete sample of all bright galaxies in their core samples, so it could not be argued that the void was due to incomplete data, as some in the West had (as it turns out, unfairly) complained of the data of Einasto and colleagues.

The Zeldovich-Einasto-Shandarin Paper

In 1982, Zeldovich, Einasto, and S. F. Shandarin used available redshift data to plot galaxies on various slices through the universe. These two-dimensional cuts through the universe showed empty cells surrounded by walls, just as two-dimensional numerical simulations by Doroshkevich had indicated. The slices were 487 million light-years by 487 million light-years and were 24 million light-years thick. If you take a two-dimensional slice through a warehouse whose volume is filled by cardboard boxes, you will find empty squares, surrounded by cardboard edges, filling your slice. Their slices looked just like cuts through an irregular Voronoi honeycomb (see Figure 4.2).

The paper emphasized that these slices favored Zeldovich's adiabatic fluctuation model over isothermal fluctuations. It was quite a tour de force. Zeldovich brought the theory. There was a detailed discussion of the Zeldovich picture of pancakes making cells, filaments forming where cell walls met, and clusters forming at the corners of the cells. Einasto brought the observations. The slices showed the filaments he had found. There was a long discussion (with several pictures) of the local Virgo Supercluster, which was a flat plate. This was described as

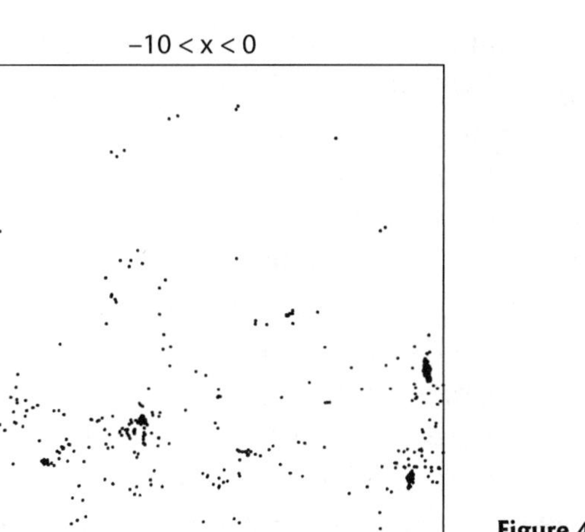

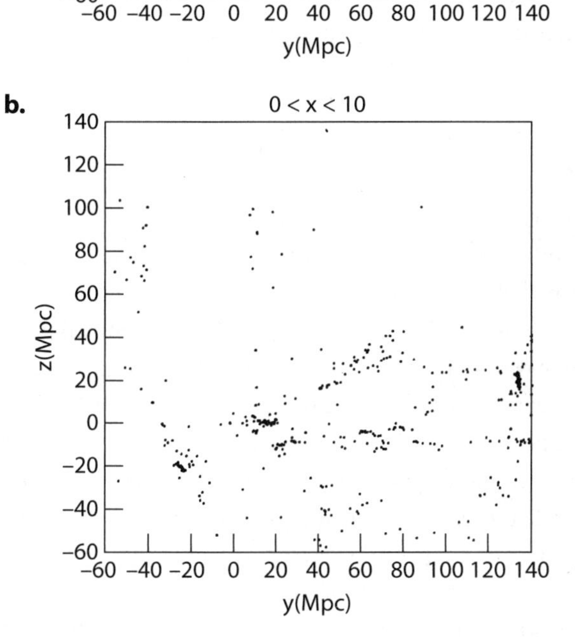

Figure 4.2. Two separate thin slices through the universe (from Zeldovich, Einasto, and Shandarin, 1982) are each 24 million light-years thick. They are shaped like giant pizza boxes, 487 million light-years square, but not very thick (not projecting much off the page). Galaxies are points, and large polygonal voids can be seen. These looked very much like slices through a Voronoi honeycomb. (Credit: From Ya. B. Zeldovich and J. Einasto, *Nature*, 300: 407, 1982; reprinted in J. Einasto, *Dark Matter and the Cosmic Web Story*, Hackensack, NJ: World Scientific, 2013, p. 225)

a wall between two empty cells. Shandarin brought simulations. These were *N*-body computer simulations based on the adiabatic picture, which turned out to fit the observations much better than simulations based on the Peebles-Soniera hierarchical clustering model. The test they used was the size of the largest connected structures based on a "friends-of-friends" approach. Draw a sphere around each galaxy of a certain radius and link it to each "friend" galaxy within that radius. As the radius for friendship is increased, the structures containing all friends of friends grow larger. The Shandarin simulations showed the size of the friends-of-friends networks grew rapidly as the radius of the spheres linking them increased. Soon, the largest network spread all the way across the entire observational box. The observations followed exactly the same pattern. The high-density regions containing the galaxies looked connected across the entire universe. The Peebles-Soniera hierarchical clustering simulation, which generated clusters of clusters (a meatball topology of isolated, high-density structures), had connected structures, the largest of which were not as big as those in the observations. I always took this famous paper to be the archetypal one in favor of a cell structure for the universe, the picture that Zeldovich proposed.

Simulations

In a 1983 paper, A. A. Klypin and S. F. Shandarin, working at the Keldysh Institute of Applied Mathematics in Moscow, published detailed results of a large three-dimensional computer simulation, which seemed to support the Zeldovich scenario—just like the *N*-body simulation shown in the Zeldovich-Einasto-Shandarin paper. They simulated a cube 778 million light-years on a side and used 32,768 particles to simulate the matter. The initial density fluctuations followed the scale-free form favored by Zeldovich, one that would make the universe continue to be equally lumpy on the largest scales as it expanded. But fluctuations were erased on scales smaller than one quarter of the initial box size. The fluctuations were random, but there were no fluctuations at small scales, rather like an ocean with random waves, but no waves below a certain wavelength (195 million light-years). This restriction was meant to simulate the natural lack of small-scale fluctuations expected for adiabatic fluctuations involving massive neutrinos. This was the most detailed simulation yet of the Zeldovich universe.

Klypin and Shandarin described their findings in that paper's abstract: "The structure formation begins from pancakes, then evolves to the network structure, then to formation of clumps connected by strings and afterwards to formation of large isolated clumps." Their results seemed to support the Zeldovich picture. In the final conditions of their model, corresponding to a picture taken at the present epoch in the universe, they looked for pancakes but said they were difficult to find, a problem they attributed to the likelihood of their being curved. They found it easier to see filaments of galaxies where they thought several pancakes came together. These filaments connected the clusters. If they looked in a plane perpendicular to a filament, they thought they found evidence for several pancakes meeting up.

In the abstract of their paper, they seemed to be saying that their results supported the Zeldovich pancake and honeycomb picture, but in their conclusions, they admitted that: "The regions of high density seem to form a single three-dimensional web structure. However, it is not clear from our simulations whether the honeycomb structure arises or not." They were having a hard time finding the pancakes. Still, this was just a simulation. In the *real* universe, the flat, pancakelike local Virgo supercluster separating two voids did seem to suggest a honeycomb.

Explosions?

In 1981, Jerry Ostriker and Len Cowie proposed a galaxy formation scenario based on thermonuclear explosions—like those Zeldovich worked on, but on a gargantuan scale. Suppose a galaxy were to form in the early universe before heavy elements had been cooked in stars. Star formation in such a galaxy might spawn a first generation of massive stars that would burn hydrogen in their cores into heavier elements and end their lives as supernovae. This could cause the galaxy to blow up, sending out a violent wind. The shock wave from this blast could increase the density of surrounding intergalactic gas and trigger more galaxy formation. These new galaxies could also spawn supernovae keeping the blast going. The gas from the expanding void in the center would snowplow up at the edges to create a shell of high-density material from which galaxies could form. Ostriker and Ed Bertschinger worked out the details of this scenario. If this occurred in a universe of critical density, the problem had no intrinsic scale: the shape of the underdensity in the void and the width of the snowplowed density enhancement at the edges of the void just scaled up linearly with size as the void expanded. This process would create an empty bubble with a shell of galaxies at the spherical bubble surface. As the bubbles collided, they would form a froth of bubbles, essentially similar to the Voronoi tessellation that Zeldovich had envisioned. Simultaneously, Peter Goldreich at Caltech came up with a similar paper, exploring the same general idea.

When Jerry explained their work to me, he noted that you would get similar results if you started off with a spherical region in the universe whose density was below average. It would be missing some gravitational binding energy because it was less dense but would have the same kinetic energy per galaxy due to the Hubble expansion of the universe. Gravitational binding energy is negative, so missing some of it was actually an energy *excess*. An excess of energy is like a nuclear bomb—it would act just like an explosion at the center. Having fewer galaxies than average, the underdense spherical region would be less decelerated than average by gravity. The underdense spherical region would begin to expand faster than the galaxies surrounding it and its galaxies would begin to run into the ones outside. They would snowplow outward, building up the Ostriker-Bertschinger shell of excess density outside the growing void.

This problem was exactly the inverse of the one Jim Gunn and I had studied (we examined what would happen to a spherical *overdense* region). These new papers were saying what would happen for a spherical *underdense* region. Once I understood that, I understood that for me at least, the nuclear explosions were unnecessary; I figured that gravity alone would do the job. Through the operation of gravity, density enhancements would produce clusters—meatballs in an expanding average-density universe—whereas density decrements would produce expanding voids surrounded by shells of excess density—like the bubbles in Swiss cheese. The Zeldovich-Einasto-Shandarin paper discussed the Ostriker-Cowie explosion scenario. They said it might be differentiated from the Zeldovich pancake picture by looking at the metal abundance of any gas detected in the voids. If the Zeldovich picture were correct, the gas in voids should be primordial—hydrogen and helium with no heavier elements. But if Ostriker and Cowie were correct, then any gas in voids would be enriched in heavier elements since it had all been through the nuclear blast wave where hydrogen and helium were processed through stars. Both scenarios seemed to result in a honeycomb universe.

Recollections of the Controversy

When Jaan Einasto came from Estonia to Princeton to visit in the spring of 2014, he and Peebles and I sat and reminisced over one long afternoon. Einasto and I talked again the next day—it was great to see him again after so many years. I related that I had always reminded people of his slices (in Figure 4.2) when similar observational results appeared later. Einasto related that in the Zeldovich-Einasto-Shandarin paper, he had to be diplomatic in what he wrote because Zeldovich was the first author. He pointed out to me something I had not noticed at the time, that the two slices shown in Figure 4.2 are actually parallel, like two adjacent slices of bread in a loaf. The reader was supposed to notice that the two slices were *different*, which meant that one was really looking at narrow filaments. Einasto was always finding filaments. He, like Shandarin and Klypin, was finding it difficult to discover pancakes. There was one sentence in the Zeldovich-Einasto-Shandarin paper that referred to the pancakes in the observations as "controversial." Einasto wrote many papers where he referred to the cell structure of the universe. He told me in 2014 that in those days he had conceived of the cells as open, the way a jungle gym made of rods can create cubic cells. I was surprised, because the word *cells* always seemed to me to imply a closed honeycomb structure. I looked up the definition of cell, and there were two meanings: (1) an enclosed room such as the cell of a monk or a prisoner's cell, and (2) a cell in biology, containing a nucleus enclosed by a cell membrane. In both cases, a volume completely surrounded by walls was implied. The word *cell* turned out to be a charged word, reminding me of the time Italian astronomer Giovanni Schiaparelli announced that he had found "canali," or channels on Mars. Percival Lowell in America heard this as *canals*—leading him to thoughts of Martian civilizations. To me, cells were closed, surrounded by walls—Zeldovich pancakes. But, allowing for his looser definition of cells, Einasto's observations were actually not as monolithic in their support of Zeldovich pancakes as I had thought.

In our 2014 conversation, Peebles said that what originally convinced him that galaxies (not clusters) formed first was that their stellar populations were very old, while he could see clusters continuing to form today by infall. Einasto said that the low individual velocities of galaxies were what first convinced him that the galaxies must have formed on the large-scale structures we see today. In the age of the universe, they would not have had time to travel far from their birthplaces relative to the size of the structures they are living on today—so the structures must have already been there.

By 1983 the overarching battle lines in cosmology were set. Would the American school led by Peebles triumph, or would Zeldovich win in the end? Would it be meatballs or honeycombs? A newly discovered theory for the origin of the universe—inflation—would provide a clue.

Chapter 5

Inflation

The Big Bang could explain many things. It was the natural result of extrapolating the expanding universe seen today back to earlier times. George Gamow and his students Herman and Alpher—through their successful prediction of the cosmic microwave background and the abundances of the light elements—showed that the standard Big Bang model was accurate back to an epoch just 3 minutes from the beginning (the epoch when the light elements were formed). Before that, one could not say if the Big Bang model was accurate. Boldly extrapolating the Big Bang model back to the very beginning, at $t = 0$, would give a singularity, a state of infinite density and curvature. Before that point, even general relativity could not extrapolate. Space and time were both created at $t = 0$.

But this leaves unanswered questions. What started the Big Bang at $t = 0$? Why is the universe so large? And why is the expansion so uniform? At time $t = 1$ second (after the Big Bang), the temperature was about 10^{10} K. That is about 3.7 billion times hotter than today's temperature of 2.725 K—the wavelengths of the microwave background photons were compressed by a factor of about 3.7 billion back at $t = 1$ second. Thus at $t = 1$ second after the Big Bang, the universe was a factor of about 3.7 billion times smaller than it is today. Today we can see out to a distance of about 13.8 billion light-years, because the universe is 13.8 billion years old. When we look out to a distance of 13.8 billion light-years, we are looking back to the cosmic microwave background that was emitted 13.8 billion years ago, just 380,000 years after the Big

Bang. We can see everything within a radius of 13.8 billion light-years—that constitutes the visible universe, and we cannot see beyond it. At $t = 1$ second, the entire region we can see today would have a radius of only about 3.7 light-years, that is, 13.8 billion light-years shrunk by a factor of 3.7 billion. But if you lived at the epoch of $t = 1$ second (disregarding that it would be much too hot for you to live there), then you could see out to a distance of only 1 light-second.

Consequently, the universe we can see today was not *causally connected* initially. What does this mean and why is it important? At $t =$ 1 second, only regions about 1 light-second in size will have had time to communicate with each other and equilibrate in temperature. Why should regions further apart than 1 light-second be at the same temperature at the epoch 1 second after the Big Bang? They have never met. They have never said "hello" by exchanging photons. But the Big Bang universe at 1 second must be uniform out to a radius of 3.7 light-years from our position or else the cosmic microwave background would not be uniform today (i.e., essentially at the same temperature no matter which direction we look). So how did the Big Bang universe get to be so uniform over large scales? There was simply not enough time after the Big Bang for different regions to *get* into causal contact.

We can draw a spacetime diagram of the expansion of the universe in the Big Bang model (Figure 5.1).

In this diagram we show time as the vertical coordinate and a single dimension of space as the horizontal coordinate. We are leaving out the other two spatial dimensions to make the result easier to visualize on a simple graph. Worldlines of particles are shown as curved black lines, tracing the paths of the particles with time. In the center is the worldline of a particle at our location. By the principles of special relativity, we are able to think of ourselves as being "at rest." Thus, the particle at our location has a worldline that is shown as a straight vertical line, not moving in space but moving up in time toward the future. It is not moving relative to us. Light travels upward and to the left or right at 45 degrees in this diagram. Particles to the right of us have worldlines that are moving to the right with time (relative to us) because of the expansion of the universe. Particles to the left have worldlines moving to the left with time (relative to us) because of the expansion of the universe. As time goes forward (toward the top of the diagram), the worldlines

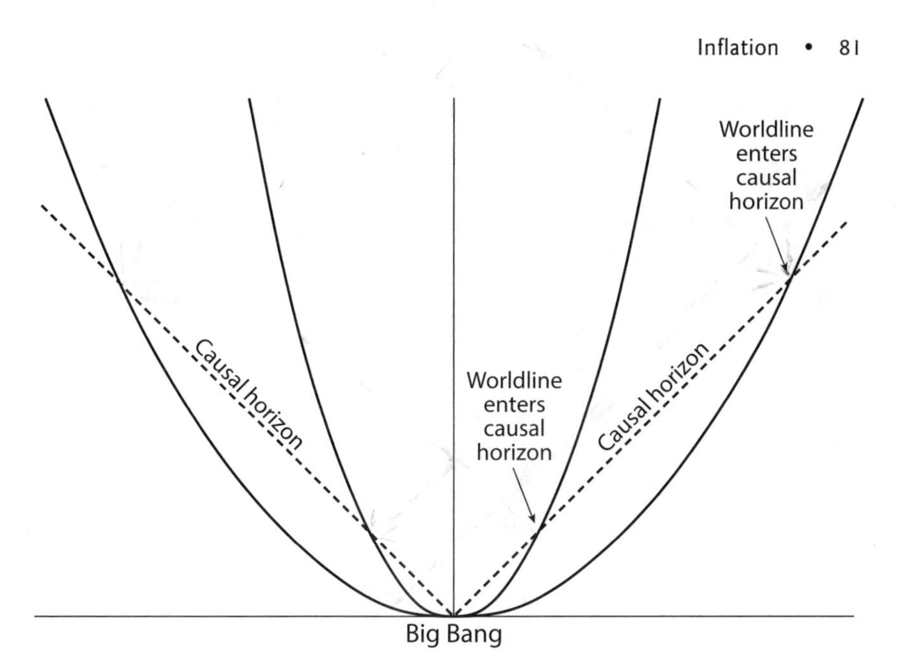

Figure 5.1. Spacetime diagram of the Big Bang model shows one dimension of space horizontally and the dimension of time vertically, future toward the top. The universe begins with a Big Bang. Curved lines are trajectories of particles (worldlines). Light beams move upward to the left or right at a 45° angle. Dashed lines show causal horizon for the central particle whose worldline is vertical. The space between the central particle and the other particles is initially stretching faster than the speed of light. Because of gravitational attraction, the worldlines of the particles are decelerating with time and eventually cross within the causal horizon, where they begin to interact with the central particle. Zeldovich hypothesized fractional fluctuations in density just coming inside the causal horizon of the central observer should be of constant amplitude as a function of time. (Credit: J. Richard Gott)

spread out. The Big Bang itself is at the bottom where all the worldlines converge. There is no time before that. In the beginning, space between two distant worldlines is separating faster than the speed of light. This is allowed and indeed predicted by general relativity. Einstein's theory of special relativity says that a rocket cannot pass you faster than the speed of light, but nothing prohibits the space *between* particles stretching faster than the speed of light. When space stretches that fast, the two particles cannot exchange photons in the time available since the Big Bang.

This diagram with its spreading worldlines is charting the expansion of space between the worldlines as a function of time. How big a region can communicate with us—the center worldline—in an elapsed time of

1 second after the Big Bang? The answer is a region of radius 1 light-second. How big a region can communicate with us in an elapsed time of 2 seconds? A region that is 2 light-seconds in radius. The boundary of the region containing worldlines with which we can communicate increases in radius linearly with time. This boundary is shown as the two dashed lines, tipped at 45°, proceeding outward at the speed of light in the left and right directions and labeled *causal horizon*. At 1 second after the Big Bang, the dashed boundary is 1 light-second away on the left and 1 light-second away on the right; by 2 seconds after the Big Bang, it will be 2 light-seconds away in each direction. This causal horizon expands linearly with time.

You will notice that worldlines distant from us lie outside the causal horizon at early times—they are initially further away than the causal horizon. Those particles cannot be seen by us until they cross the causal horizon. The worldlines of the particles to the left and right of us are not straight—straight worldlines would correspond to expansion with constant velocity, neither accelerating nor decelerating. Instead, these worldlines are all bent in a concave fashion—they are decelerating. They follow parabolas. Their recession velocities relative to us are becoming smaller and smaller with time, so the worldlines' angles of tip relative to the vertical are becoming less and less as one follows the worldlines upward in the diagram. The expansion rate of the universe is slowing down—due to gravitational attraction.

In Einstein's theory of general relativity, gravity is caused by the curvature of spacetime. So, the bending of the worldlines in Figure 5.1 is caused by the curvature of spacetime. The 1922 Friedmann universe spacetime looks like a football set up vertically ready for kickoff, as shown in Figure 1.3. Consider the seams in the football. These are lines on the football fanning out from a point at the bottom of the football (the Big Bang). These lines are as straight as you can draw on the football—geodesic paths. Because of the curvature of the football's surface, however, these seam lines are bent, and, as they cross the equator of the football, they even begin to converge, all eventually colliding at the point at the top (the Big Crunch). Seams that were originally separating are bent by the curvature of the football until they converge. This looks like gravitational attraction. Einstein's theory explains gravitational attraction by a curvature of spacetime. Genius. Friedmann's first Big Bang model, which he published in 1922, describes what would happen in a high-density universe (with $\Omega_m > 1$): the matter in the universe would cause the expansion to decelerate and eventually turn around and become a contraction.

General relativity predicts that both energy density and pressure cause space and time (spacetime) to curve. Newton figured that mass alone caused gravity. With his famous equation $E = mc^2$, Einstein proved that energy and mass can be converted into each other and are, therefore, equivalent in their gravitational effects. If Newton had known this, he probably would have agreed that energy could be attractive and that a high density of energy would be greatly attractive. But the idea that pressure is also gravitationally attractive is a new addition by Einstein. As we have discussed, at early times in the hot Big Bang model, the energy density in the universe is dominated by radiation. The energy density of radiation is equal to the number of photons per unit volume times their individual energies. It turns out that radiation itself exerts a pressure as well—photons bouncing off a solar sail on a spacecraft can push it along like a sailboat in the wind. This radiation pressure in the early universe is *one-third* of the energy density of the radiation (in appropriate units where the velocity of light is 1, or *unity*). This pressure operates in each of *three* spatial directions (width, depth, and height), making the total gravitational attraction of this pressure ($\frac{1}{3} \times 3 = 1$), or *equal* to that of its energy density. Therefore, thermal radiation gravitationally decelerates the worldlines of particles participating in the expansion twice as much as Newton could have predicted.[1] If one plugs the proper values into Einstein's equations, one can find out exactly *how much* the worldlines are being decelerated in the early universe. The distance of the worldlines outward from the center goes up like the square root of the time—they are slowing down with time as depicted in Figure 5.1.

As we have noted, the worldlines are initially expanding away from the central worldline faster than the speed of light, putting them initially outside the causal horizon. As they slow down, they cross the causal horizon boundary. Then they can exchange photons with the vertical worldline in the center. They are in *causal contact* after they cross the dashed line. They come into view as seen by us: they "say hello." In the standard Big Bang model, new particle worldlines are always coming inside the causal horizon and saying hello. Since we have not had time to send them a signal and get it back from them by the time they cross the line, it is somewhat miraculous that they should be at the same temperature as the equivalent worldlines on the opposite side. We should expect the universe to look nonuniform. Normally, if we encounter two regions at the same temperature, the simplest explanation is that they have interacted and have come into thermal equilibrium with each other. But we have no such possibility here. There is not enough time to establish thermal equilibrium. Yet when corresponding worldlines on opposite sides of us first come into view, they are already at the same temperature, even though they have never had time to communicate with each other.

When we look at the cosmic microwave background radiation, we are looking back 13.8 billion years to a time when the universe was only 380,000 years old. If we look in two opposite directions in the sky, we are looking at two different regions separated by 84,000,000 light-years, which could be influenced by stuff only 380,000 light-years away. The regions in the cosmic microwave background that have had time to communicate with each other and be in thermal equilibrium by the epoch of recombination are only 1.16° in radius—just over 4 times the radius of the full moon. This should make the cosmic microwave background in the sky in different directions very nonuniform and random on angular scales larger than 1.16°. But we observe that it is equal to 2.725 K all over the whole sky to an accuracy of one part in 10^5, disagreeing with what a simple application of the Big Bang model would suggest. We could postulate that the Big Bang is somehow uniform, by some principle we don't understand. A uniform cosmic microwave background measuring exactly 2.725 K in all directions would give us all those nice nucleosynthesis results from the early universe (see the section on dark matter in Chapter 2). But a perfectly uniform universe would *stay* perfectly uniform and no galaxies would form—no stars, no planets, no us. That's no good either. We need small fluctuations on top of an overall uniformity. (And, of course, we do observe such small fluctuations on the order of one part in 10^5 in the microwave background.)

There was a Depression era saying: We could have some ham and eggs for breakfast. If we had some ham. If we had some eggs. Likewise, we could make galaxies and the observed uniform cosmic microwave background—if we had a uniform universe, if we had some fluctuations to boot! But, in the 1970s, we knew neither how to achieve a uniform universe nor why it should have fluctuations on top of that. We needed ham and eggs but had neither.

Zeldovich Scale-Invariant Fluctuations

Zeldovich's solution to this problem was to solve it by fiat in the simplest way possible: first he would demand uniformity, then simply add fluctuations. As new worldlines continued to cross inside the causal horizon and say hello, Zeldovich gave them a mean temperature, plus or minus a standard fluctuation amplitude that was one part in 10^4 independent of time. This amplitude of 10^{-4} would then be a constant of nature. It was (1) small enough to provide the degree of uniformity to give Gamow's results for nucleosynthesis, (2) large enough to explain galaxy formation, and (3) small enough not to be embarrassing for the observations of the cosmic microwave background, which by that time (in the 1970s) demanded that the fluctuations on the scale of 13.8 billion light-years be on the order of 10^{-4} or less. It was just a hypothesis, but one that could at least be tested. This is sometimes called the *Harrison-Zeldovich scale-invariant spectrum* of fluctuations since British astronomer Edward Robert Harrison had proposed the same idea in the West.

Martin Rees and I liked the idea. It predicted more power on large scales than a Poisson distribution of galaxy seeds—the idea that Peebles was proposing. Greater power on large scales would mean a larger distribution of cluster sizes in agreement with the observations. Zeldovich's idea seemed simple and consistent with the observations. But it was arbitrary on two fronts, postulating uniformity without a cause and then constant fluctuations, also without explanation. Was there a theory that could provide these explanations?

Inflation to the Rescue

In 1981, Alan Guth at Stanford proposed a theory he called *inflation* to explain the uniformity of the universe and its enormous size. We are used to thinking that empty space (the vacuum) should have a zero energy density. But Guth proposed that in the early universe the laws of physics were different and there was a different *vacuum state* giving empty space a high amount of energy per cubic centimeter. In order that all observers traveling in rocket ships at different speeds would observe the same vacuum energy density, it was necessary by the logic of Einstein's special relativity that the vacuum should also have a *negative* pressure equal in magnitude to the energy density. In this case, the vacuum state has no preferred standard of rest, just as Einstein supposed in special relativity.[2] This vacuum state looks exactly like the cosmological constant term Einstein added to his equations (in 1917) to produce a static universe, and it is gravitationally repulsive, because in Einstein's theory of general relativity, pressure gravitates as well as energy density. The negative pressure operates in the three spatial directions (width, breadth, depth) and, because it is *negative*, it has a *negative*, or *repulsive*, gravitational effect three times as large as the gravitational attraction of the energy density. That means that the overall effect of this positive energy-density vacuum state is gravitationally repulsive. Start the universe off in a static configuration and it will begin to expand; its geometry will be like the top half of de Sitter spacetime (shown in Figure 1.2). As its expansion approaches the speed of light, the clocks on individual particles will slow down due to Einstein's theory of relativity. Measured by these ever-slowing clocks, the expansion becomes exponential as a function of time. The separation of worldlines with time doubles every 10^{-38} seconds, as measured by the particles. This creates a sequence that increases rapidly: 1, 2, 4, 8, 16, 32, 64, 128, 256, 512, 1,024, ... After 10 doubling times, or 10^{-37} seconds, the universe will have increased in size by a factor of a thousand. After another 10 doubling times, or 2×10^{-37} seconds from the start, the universe will be a million times larger. After a thousand doubling times, or 10^{-35} seconds, the universe will be $2^{1,000}$ times larger or about 10^{300} times larger. The universe is going to get very big, very fast. This is going to account for the enormity of the universe.

Pick an epoch slightly after the beginning of the Big Bang model and cut off the part earlier than that and throw it away. (Cut off the bottom of the football in Figure 1.3, like cutting off the tip of a cigar.) Replace this with a short epoch of exponentially expanding empty space with a high energy density and negative pressure—just like the one we've been considering. This inflationary epoch looks like a tiny funnel (like the top half of Figure 1.2) and can add a little extra time—say, 10^{-35} seconds—onto the Big Bang model at the beginning. This will give different regions extra time to come into causal contact. The worldlines start off closer than 10^{-35} light-seconds apart so they have time to get into contact and into equilibrium in the extra 10^{-35} seconds of the inflationary epoch. Then the inflation will spread these regions apart by a factor of $2^{1,000}$, or about 10^{300} (pushing them outside the causal horizon). This factor of 10^{300} is so large that all the distant regions we can see today would have been in causal contact in the early inflationary phase, and we would expect them to be at the same temperature. At the end of the inflationary stage you have a very large, but uniform, universe.

What happens then? Guth supposed that the laws of physics changed at that point and that the energy in the vacuum state decayed into ordinary thermal radiation, the stuff of the standard Big Bang model. The energy density at the moment of change would remain constant, but the pressure would go from being equal to −1 times the energy density to being equal to +⅓ times the energy density (the value associated with thermal radiation). This thermal radiation is gravitationally attractive and so the expansion starts to *decelerate* in the Big Bang phase of the model. This gives an *inflationary-Big Bang model*. as illustrated in Figure 5.2.

Compare this with Figure 5.1. An early inflationary epoch has been added to the bottom portion of the diagram to provide extra time for causal communication.

In the inflationary epoch (a funnel-like shape), the worldlines are accelerating outward as time proceeds upward; once the worldlines pass into the Big Bang epoch, where the universe is dominated by thermal radiation, the worldlines are decelerated (a football shape), just as in the standard Big Bang model. The dashed lines show the causal horizon for the worldline in the center (us). In the Big Bang epoch, the causal horizon is moving outward with time as depicted before, but below this, during the inflationary epoch, the causal horizon is at a constant distance from our central vertical worldline. How far away from us is it? Approximately 10^{-38} light-seconds away. Why? Because the doubling time for the inflationary expansion is 10^{-38} seconds. Early on, we are in causal contact with a distant worldline, when it is inside the causal horizon. But after it crosses the causal horizon, we lose sight of it. Its distance from us is doubling every 10^{-38} seconds due to the stretching of space

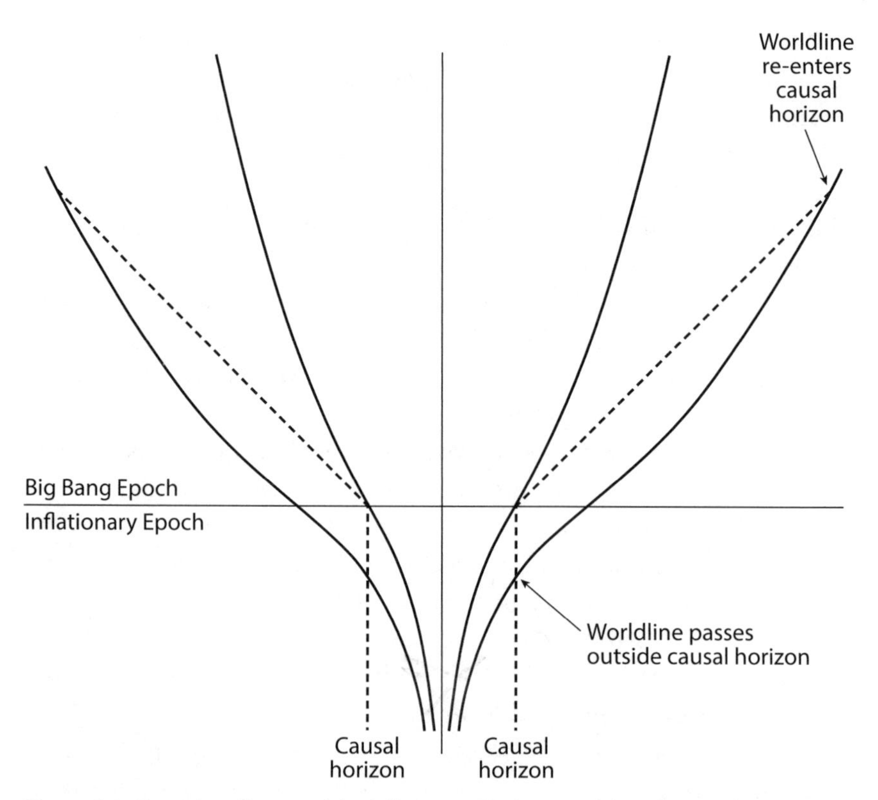

Figure 5.2. Spacetime diagram of the Inflationary Big Bang model (with space horizontal and time vertical as before). It starts with an inflationary epoch at the beginning, during which the universe doubles in size in equal intervals of time: 1, 2, 4, 8, ... Worldlines of the other particles accelerate away from the central particle due to the gravitationally repulsive effects of the vacuum. During the inflationary epoch, the causal horizon for the central particle stays at a constant distance. Particles originally in causal contact pass outside the causal horizon as the space between them and the central particle begins to stretch faster than the speed of light. Inflation ends as the energy in the vacuum is dumped into thermal radiation. After that, the worldlines decelerate, eventually coming back inside the causal horizon during the Big Bang epoch (as in Figure 5.1). (Credit: J. Richard Gott)

between it and us. Just as it crosses outside the inflationary causal horizon, its distance from us is 10^{-38} light-seconds. In another 10^{-38} seconds, its distance will double to 2×10^{-38} light-seconds—that is, in a time of 10^{-38} seconds more, it will move a distance of 10^{-38} light-seconds further away due to the stretching of space. The space between it and us is thus stretching at that point at exactly the speed of light, and a light beam traveling from it to us will just barely reach us. In another 10^{-38} seconds, it will be 4×10^{-38} light-seconds away, and the space between us will be stretching at twice the speed of light. Once a particle passes the causal boundary at 10^{-38} light-seconds away, we lose sight of it: we say goodbye to it.

But after the worldline of the particle passes into the Big Bang epoch, it starts decelerating, slowing down until it eventually crosses back through the dashed causal horizon line, which begins expanding linearly with time during the Big Bang epoch. Then we can say hello again to the particle. At that point it can report on its adventures when it was beyond the causal horizon. In the simple Big Bang model, we would have been saying hello for the first time to the particle at this point. In the inflationary Big Bang model, we are saying hello *again* because we already got acquainted in the early inflationary phase—we expect to have established equilibrium with it already. When we say hello again, we expect it to be at the same temperature as the symmetric worldline on the other side of the diagram, which is also crossing through the dashed line on the other side and also saying hello again. Therefore, we expect to see thermal radiation at the same temperature on one side of the sky as on the opposite side of the sky. We are saying hello again to two old friends whom we have met and equilibrated with in the past rather than saying hello for the first time to two strangers (as would occur in the standard Big Bang model). This inflationary process explained how the microwave background radiation we see could be so uniform. This was a very important advance.

Quantum Fluctuations

But there is more. One should expect random quantum fluctuations in the original region of 10^{-38} light-seconds across because of the uncertainty principle. Since the velocity of light is 3×10^{10} centimeters/second, 10^{-38} light-seconds is just 3×10^{-28} centimeters. That is a tiny, subatomic scale. We expect quantum fluctuations on this microscopic scale. Heisenberg's uncertainty principle tells us that on a timescale of 10^{-38} seconds, we can measure an energy only with a limited accuracy.

In a brilliant 1982 paper, Stephen Hawking figured out that inflation would automatically produce fluctuations that would obey the Zeldovich scale-invariant hypothesis. Definitely a Eureka moment! This was later elucidated in more detail in 1983 by James Bardeen (son of John Bardeen, who discovered the transistor) who showed, along with his colleagues, that if these quantum fluctuations leave the causal horizon (and say goodbye) with a certain amplitude, they will retain that *same* amplitude when they reenter the causal horizon and say hello again. It is a result that follows from general relativity theory. In the diagram, the most distant pair of particles leave the causal horizon first: when they leave, they are separated from us by 10^{-38} light-seconds. They look exactly like the nearest pair of particles when they leave the causal horizon later, also separated from us by 10^{-38} light-seconds. In each case, as they leave they carry with them fluctuations on that scale that are of the *same* constant amplitude. They each leave under similar circumstances, and when these particles come back inside the causal horizon and say hello again, they carry the *same* constant amplitude fluctuations with them as they reappear, by Bardeen's proof. Thus the fluctuations coming back into view across the causal horizon and saying hello again are of constant unchanging amplitude. This is the Zeldovich postulate! Inflation produces this result automatically.

The Hawking-Bardeen and colleagues' result may also be explained in the following way. Fluctuations outside the causal horizon decrease in amplitude during the remaining inflationary phase because inflation tends to smooth out wrinkles in the geometry of the universe by stretching them, and if curvature fluctuations go down, so do density fluctuations, because they are directly linked by Einstein's equations of gravity. These same fluctuations increase in amplitude during the Big Bang phase because of ordinary gravitational instability: denser regions experience more deceleration and become denser still relative to the average as they evolve, while less-dense regions do the opposite. That increase in amplitude during the Big Bang phase exactly cancels the decrease during the inflationary epoch.

Some caveats remain. The amplitude fluctuations have when they come back inside the causal horizon is the same as it was when they left—*except* for a small factor, which depends on the ratio of pressure to energy density obtained in the universe when they say goodbye and hello again. This factor is larger for particles saying hello again in the early Big Bang phase, which is dominated by radiation, than it is later, when the universe becomes dominated by matter. This means that the amplitude of fluctuations coming within the horizon during the radiation-dominated phase is larger than the amplitude of the fluctuations entering the horizon during the matter-dominated epoch. This effect is actually helpful for the formation of galaxies and clusters of galaxies. The universe becomes matter dominated when the universe is about 3,234 times smaller than it is today. The size of the causal horizon at this epoch, expanded by a factor of 3,234 to show the size the fluctuations would have today, is about 400 million light-years across. Thus the scales on which galaxies and clusters of galaxies are forming (i.e., smaller than 400 million light-years across today) came within the causal horizon and said hello again when the universe was still radiation dominated. So they get the benefit of this larger factor in amplitude starting off. This boosts the formation of galaxies and clusters of galaxies, while keeping the fluctuations in the cosmic background as small as what we observe today. The Harrison-Zeldovich constant-amplitude hypothesis would not give us this extra saving grace.

Guth proposed that the inflationary vacuum state is produced by the action of a field that permeates space. One field with which we are familiar is the magnetic field. The magnetic field has a strength (a magnitude) and a direction. You can use your compass to show you the direction of the magnetic field. At every point in spacetime, there is a magnetic field strength and a direction. The kind of field proposed for inflation is a *scalar* field; it has a magnitude but points in no particular direction (rather like the recently confirmed Higgs field, which confers mass on particles). The inflationary field sets the value of the vacuum energy density. As the magnitude of the field changes from one region of spacetime to another, so does the amount of vacuum energy density (see Figure 5.3).

Imagine a landscape where the altitude represents the value of the vacuum energy density. Following work by Linde and Albrecht and Steinhardt, imagine that in the early universe the vacuum energy is high, corresponding to a position sitting high in a mountain valley, at high altitude. Put a ball there. Its horizontal position (left to right) represents the value of the field (low values on the left, high values on the

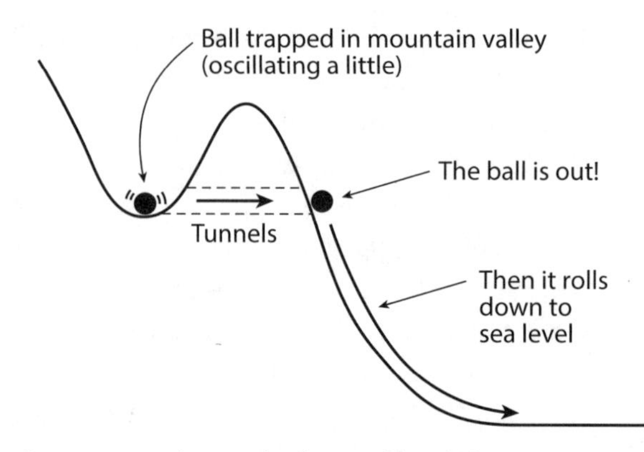

Figure 5.3. Quantum tunneling can be illustrated by a ball sitting at the bottom of a high mountain valley. Its altitude above sea level represents the value of the vacuum energy. It would stay there forever, except that, due to its quantum mechanical wave nature, it is not well localized. It has a small chance of tunneling through the mountain and finding itself suddenly outside on the slope, where it can roll down toward sea level. (Credit: J. Richard Gott, *Time Travel in Einstein's Universe*, Boston: Houghton Mifflin, 2001)

right). The high altitude of the ball sitting in the high mountain valley represents the high energy density of the vacuum in the early universe. The ball would be happy to stay there forever if it were not for *quantum tunneling*.

How does this work? Quantum tunneling is due to the wave nature of the ball. The ball is not located at a certain place: rather, its probability of being found at particular places is spread out due to its wave nature. You could consider it oscillating slightly at the bottom of the mountain valley. The ball has a small probability of suddenly finding itself completely outside of the mountain valley, somewhere along a high, gently sloping plateau, for example. If that happens, we say the ball has "tunneled" out of the mountain valley onto the plateau. Now the ball can roll slowly down the gently sloping plateau toward sea level. Finally, the ball falls down to near sea level—the vacuum state the universe is in today.

As the ball rolls down the hill, the value of the field changes (i.e., the horizontal position of the ball changes) and the vacuum energy density (its altitude) decreases. Finally, as it reaches the bottom of a valley near sea level, the ball rolls back and forth and dissipates its kinetic energy in the form of thermal radiation and comes to rest at the bottom of a near sea-level valley. This marks the transition to the Big Bang phase, where the vacuum energy becomes low and the universe becomes filled with thermal radiation. As we reach the end of the inflationary epoch, there is significant kinetic energy in the field (rolling down the hill), and this raises the pressure above −1 times the energy density. This changes the factor by which the amplitude of the fluctuations change between when they say goodbye and hello again. Also, as the ball rolls down the hill, the energy density in the vacuum slowly changes with time in the inflationary phase, causing the doubling time for the exponential expansion to slowly lengthen. These effects create a small tip in the fluctuation spectrum as a function of scale relative to the constant value predicted by Zeldovich. As we will see in Chapter 10, a tip of just the expected size has now actually been observed in the cosmic microwave background data. Zeldovich gave us scale-invariant fluctuations by fiat. Inflation gave us *almost-scale-invariant fluctuations* that were even better: deduced from physics and in better agreement with the observational data.

In principle, inflation gave us both the ham and the eggs. But it presented its own problem immediately, one which Guth himself pointed out. Guth needed the transition between the inflationary epoch and the Big Bang epoch to occur all at once: he needed the high-density vacuum state of the early universe to decay all over the universe simultaneously and deposit its energy in thermal radiation. This was a phase transition, going from a high-density vacuum state to thermal radiation—analogous to boiling a pot of water on the stove and having the liquid water turn to steam all at once. It's not likely to happen that way; it's likely to form bubbles, with hot steam in each of the bubbles. But allowing such bubbles in the model would destroy the very uniformity of the universe that inflation was trying to explain. Guth found that bubbles would form by quantum tunneling, a phenomenon investigated earlier by Sidney Coleman. Low-density bubbles would just randomly appear and start expanding. According to Coleman's formulation, the bubbles would tunnel directly to sea level. Unlike Figure 5.4, the tunnel would slope diagonally downward to emerge at sea level—a zero-energy vacuum state. These Coleman bubbles would be empty inside, containing a zero-energy-density vacuum state. The density in the inflating region is high and the bubbles are empty. The bubbles quickly begin to expand at nearly the speed of light. They have zero pressure inside and a negative pressure outside in the inflating region, so the negative pressure simply pulls the bubble wall rapidly outward. Guth found that although the bubbles expanded at nearly the speed of light, the inflating region was expanding even faster, in such a way that the bubbles did not percolate to fill the entire space. Between the empty bubbles, there was an endlessly inflating sea of high-density vacuum. What was left was not uniform. This was a conundrum.

Bubble Universes—a Multiverse

In January of 1982, I published a paper in *Nature* proposing an answer to this question. I proposed that bubbles did form in the endlessly inflating high-density vacuum sea, but that inflation continued *inside* the bubbles for a while. I proposed that once a bubble formed, it retained its high-density vacuum state for a short period; only somewhat *later* was this vacuum energy dumped within the bubble in the form of thermal radiation. My bubbles were not empty. We simply lived *inside* one of the bubbles. From an observation point inside one of the bubbles, the bubble looked uniform and at late times like an $\Omega_0 < 1$ Friedmann Big Bang model. The bubble looked like an ever-expanding, inflationary Big Bang universe. We, looking out in space and back in time, could see only our own bubble and the smooth inflating sea that produced it. We could not see the other bubbles. The inflation occurring within our bubble after its formation by quantum tunneling was large, say, 100 or more doubling times. This would create uniformity within our bubble, just as in Figure 5.2.

I proposed that we lived inside a bubble universe and that our universe was just one of many bubbles. Our bubble universe would continue to expand forever. Meanwhile, outside our universe, other bubble universes would continue to form in the endlessly inflating, high-density sea (see Figure 5.4). I proposed that inflation could form an infinite number of such bubble universes. When each bubble formed, it would effectively start a clock in that bubble, which after a specified amount of time, would cause a decay of the inflating high-density state into thermal radiation.

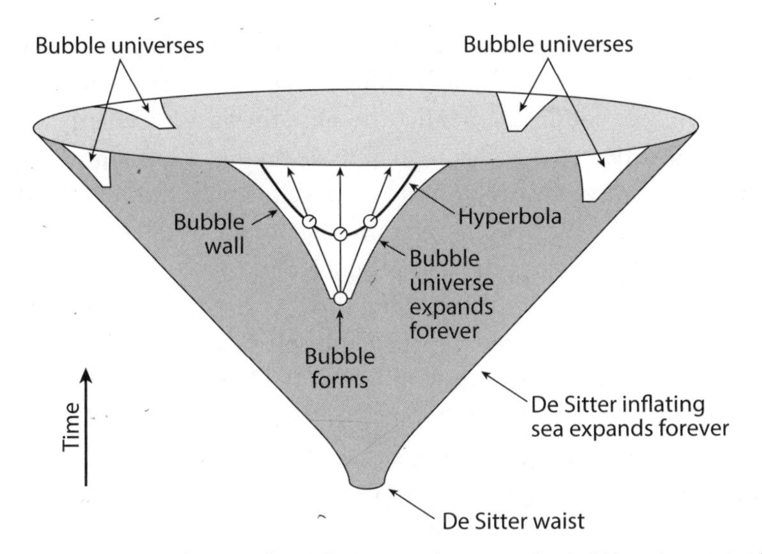

Figure 5.4. Spacetime diagram of an inflationary universe creating bubble universes (with time vertical). Start with a tiny region containing vacuum energy (at the deSitter waist at bottom), and its repulsive gravitational effects will start the space expanding faster and faster, continuing forever. Bubbles of lower-density vacuum energy can form in this inflating sea. These expand, forming individual bubble universes. Inside a bubble, we see worldlines of particles accelerating away from each other as inflation continues for a while within the bubble, until the vacuum energy is dumped in the form of thermal radiation (where the little circular alarm clocks go off). After that, each bubble enters an expanding Big Bang phase. This creates a multiverse of bubble universes that expand forever. (Credit: J. Richard Gott, *Time Travel in Einstein's Universe*, Boston: Houghton Mifflin, 2001)

This transition epoch is causally related to the bubble's formation event. No synchronization of non-causally-connected regions is required (as was the case in the Big Bang model). This seemed to answer Guth's problem *and* it produced what we now call a *multiverse*. In my 1982 *Nature* paper I said, "Thus we can see the formation of our Universe as a quantum tunneling event; our Universe is one of the normal vacuum bubbles."

A short time after my paper appeared, independent papers by Andrei Linde and by Andreas Albrecht and Paul Steinhardt appeared which, proposed detailed particle physics scenarios producing exactly the geometry I had suggested. As in Figure 5.3, they proposed that initially the quantum vacuum state would be like a ball sitting in a high mountain valley. The ball would be happy to stay there forever, were it not for quantum tunneling. When the ball tunnels out, that forms a bubble in which the vacuum state inside the bubble is still at a location high on the hill. The bubble continues to inflate because the vacuum energy density is still high—the ball is still at high altitude on the hill. But now the ball can roll slowly down the hill toward sea level. Finally, the ball falls down to near sea level—the normal vacuum state we are in today.

The kinetic energy of the ball produced by rolling down eventually gets converted into thermal radiation, and the solution starts to look like a Big Bang. What about the time taken to roll down the hill? I could see that was exactly the clock I needed for my bubble universe model! It always takes a specific amount of time to roll down that hill, as I described at the 1984 *Inner Space/Outer Space* conference (see Gott 1986). Tom Statler and I wrote a paper on this in 1984 showing how the fact that we had observed only our own bubble—and that, so far, we had seen no collisions with other bubble universes—allowed us to put an upper limit on the bubble-formation rate in the inflating sea, an upper limit that was safely above the low rate expected from quantum tunneling.

I was quite ready to accept Guth's 1981 paradigm shift and propose a solution to its central problem because I had been introduced to similar "inflationary" ideas earlier at a summer school for cosmology in the French Alps in 1979. I took my bride Lucy there, where we enjoyed magnificent mountain views. One day, the speaker was François Englert, who explained papers he had published with his colleagues R. Brout and E. Gunzig in 1978, and with Brout and P. Spindel in 1979. They proposed a different inflationary scenario. They were trying to create an inflating bubble from a flat empty spacetime. Inflation went on inside the bubble for a while and then turned off after a time, turning into a Big Bang model. I was sitting next to Martin Rees in the audience and I remember leaning over and saying to him, "This is very important!" Englert would later receive the Nobel Prize for his work on the Higgs field, leading to the discovery of the Higgs boson. He was indeed worth listening to. Nevertheless, the model of Englert and his colleagues seemed counterintuitive since it started with a zero-density universe and had a positive-density bubble form inside it. Still, I was impressed and remembered the talk. In my 1982 paper, I would simply start with a positive-density inflating space (de Sitter space) and make many lower-density universes inside it via quantum tunneling; this would involve always rolling downhill. Meanwhile, in 1980 Soviet astrophysicist Alexei Starobinsky, while on a visit to Cambridge University, had published a paper proposing yet a different brand of inflation, which simply started with an inflating de Sitter space, which decayed into a Big Bang model. He had made the transition occur all at once (as Guth did) and didn't address the problem of bubble formation that Guth recognized. Starobinsky had been Zeldovich's student. Again, because of the Cold War, his paper was not much noticed in the West. Another paper from 1980, published by D. Kazanas in the *Astrophysical Journal*, was titled "Dynamics of the Universe and Spontaneous Symmetry Breaking"; it proposed an exponential expansion phase in the early universe based on the Higgs field. So inflating states in the early universe were in the air. But these papers were not being noticed. They received a total of only seven citations prior to 1981. When Guth discovered an inflationary model independently and published it in January of 1981, it got wide and immediate attention. He was invited to give talks all around the United States, and managed, like Copernicus, to convince people of this new world view. It received more than ninety citations in 1981 and 1982. I published my paper proposing forming bubble universes in an inflating sea via quantum tunneling in January of 1982. Linde published his paper on "new inflation" in February of 1982, and Albrecht and Steinhardt's independent paper was published in April of 1982. A paper by Stephen Hawking on inflation producing the Harrison-Zeldovich scale-invariant fluctuation spectrum and adopting the bubble universe idea was published in September of 1982. Its title was "The Development of Irregularities in a Single Bubble Inflationary Universe." It referenced Starobinsky, Guth, Bill Press, Linde, Hawking and Moss, Albrecht and Steinhardt, Gibbons and Hawking, Harrison, and Zeldovich, and me. I'm proud to be among those referenced in this important paper.

In Figure 5.4 the multiverse begins as a tiny inflating region at the de Sitter waist at the bottom. This represents a 3-sphere universe with a circumference perhaps as small as 2×10^{-27} cm. This is very small and Alex Vilenkin (1982) and Hartle and Hawking (1983) have speculated that it may simply pop into existence via quantum tunneling from nothing (or a quantum state of zero size). This then expands forever, as one goes upward in the diagram, spawning an infinite number of bubble universes, each of which becomes infinite in size. Each bubble universe would naturally have the desired Harrison-Zeldovich scale-invariant fluctuation spectrum. It's rather amazing that, starting from nothing, or nearly nothing, inflation is capable of producing a multiverse containing an infinite number of bubble universes, each infinite in extent, expanding forever, with just the desired form of density fluctuations.

Production of Adiabatic Fluctuations

Werner Heisenberg's uncertainty principle showed that one cannot measure the position of a particle and its velocity with arbitrary accuracy. It is one of the foundation stones of quantum mechanics. The uncertainty principle also implies that the energy or value of a field will be uncertain by a certain amount due to the short timescale available for measurement in the early inflationary universe. This ensures there are random quantum fluctuations in the position of the ball (the value of the field). The position of the ball is thus "spread out" into a bell-shaped, or *Gaussian*, distribution. It gives the surprising result that, as the ball rolls down the hill, there are some regions of space where the position of the ball along its path is slightly ahead and other regions where it is slightly behind! In some regions the ball will reach bottom slightly sooner. The regions where the ball reaches bottom sooner will convert the vacuum energy to thermal radiation at that point, whereas other regions remain a while longer in the high-density vacuum state (still rolling down the hill).

Once the first region has converted its vacuum energy into thermal radiation, this radiation will lose energy density as the universe continues to expand, as the photons thin out and redshift. Thus, the first region where the ball reaches bottom early loses energy density rapidly after that, as its thermal radiation thins out, while the energy density remains high in a "neighboring" region where the ball, still rolling down the hill, continues to inflate, retaining its high density a bit longer. When *that* region finally converts into thermal radiation, it will have a higher density than the region nearby whose thermal radiation has already lost energy due to expansion. This produces small random variations in the amount of thermal radiation from place to place after the Big Bang phase has been reached.

Thus, this process creates regions of slightly hotter, denser thermal radiation, and other regions with slightly colder, less dense thermal radiation at the start of the Big Bang epoch with a bell-shaped, or Gaussian, distribution, having a symmetric, random distribution of positive and negative density fluctuations about the mean. These are exactly the *adiabatic fluctuations* Zeldovich wanted all along. Initially this thermal radiation is very hot, containing not only photons but all elementary particles and their antiparticles as well. We expect that asymmetric decay of heavy particles present in the early thermal radiation could lead to the excess of matter over antimatter seen in the universe today. Regions that initially contained more thermal radiation would be also be expected to contain more ordinary matter particles and dark matter particles as well.

In 1983 Linde would propose a theory of *chaotic inflation*, in which, in the early universe, the ball simply starts off on the *slope* of the hill. Instead of having just tunneled out of a mountain valley, it could have gotten up there by a rare quantum fluctuation. In quantum theory, particles are associated with fields (e.g., the photon is the particle associated with the electromagnetic field, while the recently discovered massive Higgs particle is the particle associated with the Higgs field). If the field responsible for inflation had a high-mass particle associated with it, the shape of the hill would be very simple, and the bottom of the valley near sea level would have a simple parabolic form—like the bottom of a bowl.

Linde's chaotic inflation could, like the bubble universe model, also lead to a multiverse. As the balls rolls slowly down the hill, the universe inflates, doubling in size again and again. When it doubles in size, its volume goes up by a factor of 8. So now you have 8 independent volumes that have passed out of causal contact with each other, and all are inflating at approximately the same density. The uncertainty principle guarantees that some of these 8 regions will have slightly higher vacuum energy than others; these will inflate faster, doubling in size more quickly. It's as if people had more children if they lived at higher altitude. In such a world, after a while, almost everyone would be living in the mountains. Most of the volume in the universe will end up in the mountains, inflating very rapidly. Occasionally (as occurs with bubble universes), a pocket will occur that rolls all the way down to near sea level. This will create a *pocket universe*. Thus, pocket universes are continually forming in an inflating sea—a multiverse.

There is a general consensus that inflation will lead inevitably to a multiverse. At Freeman Dyson's ninetieth birthday celebration, Martin Rees told how he was once asked at a conference how strongly he believed in the multiverse. Martin said he would not bet his life on it, but he would go as far as betting his dog's life. Linde rose to say that since he had spent decades working on the multiverse idea, he had demonstrated that he would bet his life on it. Nobel Laureate Steven Weinberg quipped that he would be willing to bet Linde's life on it—and Martin Rees's dog's!

Cold Dark Matter

In 1982, Jim Peebles reenters the picture. By this time, Peebles has abandoned his idea of placing galaxy seeds at random locations and has adopted the Harrison-Zeldovich scale-invariant spectrum, except now based on Hawking's 1982 paper using the inflationary–Big Bang model. Peebles proposed that the dark matter found by Zwicky was made up of *weakly interacting massive particles* (WIMPS, which we mentioned in Chapter 2 in the last section on dark matter). (Simultaneously George Blumenthal, Heinz Pagels, and Joel Primack advanced similar ideas—citing *gravitinos*, supersymmetric partners of gravitons, as the candidate WIMPS.) These WIMPS could be produced from the initial, very hot, thermal radiation. Wherever there was a slightly higher density of thermal radiation, there would also be a slightly higher density of WIMPS. The WIMPS would be weakly interacting; in particular, they would have no electric charge and would not interact with the thermal radiation. They would thin out as the universe expanded, in exactly the same fashion as the photons in the thermal radiation had, except for one crucial difference: because the photons were losing energy as their wavelengths got stretched by the expansion—while the WIMPS and ordinary matter particles did not lose energy in this fashion—the energy density in the photons would eventually fall below the energy density in the (dark matter + ordinary matter) particles. This occurs when the universe was 3,234 times smaller than it is today and about a factor of 3 smaller than it was when recombination occurred.

This is called the *cold dark matter* model. The dark matter is made of massive particles that do not have large random velocities, so we say they are cold, and having no large random velocities, they produce no appreciable pressure to resist gravitational collapse. Fluctuations in the cold dark matter on galaxy scales start to grow the moment they come within the horizon during the radiation-dominated epoch. They grow slowly at first because the thermal radiation makes up most of the energy density of the universe, dominating the gravity. But they begin to grow faster after the universe becomes matter dominated, when the universe is about 70,000 years old. Slightly denser regions get denser still, relative to the background, as they draw in additional particles by their gravitational attraction. Less-dense regions thin out more than average. The amplitude of the fluctuations in the cold dark matter continues to grow. The ordinary matter is still ionized and coupled to the thermal radiation, whose high pressure makes it stiff so that fluctuations in the ordinary matter can't start growing until recombination when the universe is 380,000 years old. This allows the density fluctuations due to cold dark matter to get a big head start on growth, a feature of the model that is very helpful in making galaxies. Cold dark matter fluctuations on smaller scales (such as galaxy scales) come into view inside the causal horizon earlier and get more of a head start on growth. Because these WIMP particles are very massive, they have slow velocities and do not leak out of even small, high-density regions. Galaxy-scale fluctuations in cold dark matter do exist and do not get erased.

Cold dark matter density fluctuations can form galaxies directly by gravitational instability. The normal matter particles would not cluster at first, being coupled to the microwave background radiation, but when freed from it at the epoch of recombination, they would fall into the already-forming halos of dark matter. Once formed, galaxies can cluster in the fashion Peebles had supposed. These cold dark matter fluctuations provide more power on large scales than Poisson galaxy seeds would have had. This would explain the wide variation in galaxy cluster sizes that Ed Turner and I had found, which required more power at large scales than Poisson galaxy seeds could deliver. The cold dark matter model could now make the hierarchy of galaxy clustering that Peebles envisioned.

Zeldovich had not used cold dark matter, only heavy neutrinos and ordinary matter. The neutrinos had a known number density and could not be very massive without violating constraints on the total density in the universe. That meant that the neutrinos had to be warm, moving around at fairly high speeds. Thus, they could escape from small structures, stopping those small structures (galaxies) from ever forming directly by gravitational instability. That is why Zeldovich wanted great pancakes of matter to form first and only later fragment into galaxies. Zeldovich could not form galaxies first. Peebles, using cold dark matter, could form galaxies and then have them cluster. The initial conditions from this cold dark matter model, with fluctuations calculated from the inflationary–Big Bang model, could be put into computer simulations to watch the galaxy formation and clustering that would ensue. Adrian Melott and his colleagues (Melott et al. 1983), started doing such calculations; so did George Efstathiou (e.g., Davis, Efstathiou, Frenk, and White 1985). What sort of clustering pattern would this new model produce? The geometry of this pattern would be unexpected.

Chapter 6

A Cosmic Sponge

Did the visible universe's structural geometry most resemble meatballs or Swiss cheese? These metaphors seemed to summarize the only two possibilities. If the low-density regions of the universe were in one connected piece, then the high-density regions must be in separate pieces, like meatballs in a cosmic soup. If, on the other hand, the high-density regions were in one connected piece, the low-density voids must be in separate pieces, like the holes in Swiss cheese. Apparently it had to be one way or the other. But that was two-dimensional thinking. If one thought in 3D, another possibility emerged, one that hadn't been considered before. I knew that a third possibility existed because of a science project I had done in high school. I had discovered some peculiar arrangements of polygons that had a structural geometry similar to that found in a marine sponge. I knew some of my *spongelike polyhedrons* divided space into two equivalent and interlocking parts. A sponge is something whose insides and outsides can be the same. A marine sponge is all in one piece but has water passages percolating through it. This brings nutrients to all parts of the sponge. If I were to pour concrete into the water, let it harden, and then dissolve away the unfortunate sponge with acid, I would be left with a concrete sponge. The concrete sponge would be all in one piece, with air passages percolating through it. It would be essentially the 3D inverse of the living sponge.

In the theory of inflation, the density fluctuations in the early universe were produced by random quantum fluctuations. Random quantum fluctuations could either be positive or negative. A positive fluctuation would cause the density in that region to be slightly above average; a negative fluctuation would cause the density to be slightly below average. Over the course of 13.8 billion years, gravity would cause these fluctuations to grow. A region that was of higher-than-average density would expand more slowly than the rest of the universe, because the gravitational attraction of its different parts for each other would be greater than average. As a high-density region began to expand more slowly than the rest of the universe, it would grow denser still relative to the rest of the universe. A region that was less dense than average would in turn have less gravitational self-attraction to slow down its expansion. It would end up expanding faster than the rest of the universe, becoming even less dense relative to the rest of the universe as it expanded. Over time, the density fluctuations would grow in magnitude, as we have discussed. At some point, the overdense regions would grow to have relative densities of 1.01, while comparable underdense regions would have relative densities of 0.99. Still later, the overdense regions would grow to have relative densities of 1.1, while comparable underdense regions would have relative densities of 0.9. Positive and negative fractional density fluctuations ($\delta\rho/\rho$) would grow by equal factors. This simple process is called *growth by gravitational instability in the linear regime*. All fluctuations grow by the same linear factor. As long as we look on large scales where the density fluctuations are of order less than 1 today, we should see a pattern of overdense and underdense regions resembling the initial conditions—but just enhanced in contrast. But what should the initial conditions produced by inflation look like?

If inflation produces density fluctuations from random quantum fluctuations, then the geometry of the high- and low-density regions must be identical. That's because random fluctuations are equally likely to be positive or negative. These inflationary fluctuations start out small—think of sinusoidal waves like ocean waves crossing space in all different random directions. These waves can be generated by a random-number generator when making a computer simulation of the universe. Make such a set of waves, going in all different random directions. The phases of the waves, the positions of their crests and troughs along the direction of motion, are also random. The amplitude of the waves will be drawn from a Gaussian, or bell-shaped, distribution. This is, therefore, called a *Gaussian random-phase* distribution. It is what inflation predicts. Some regions will be above average in density and some will be below average in density when all the waves are added up. That is a perfectly good set of initial conditions. But now multiply all those random numbers by −1. That will turn all the wave crests into troughs and the troughs into crests. It will turn all the high-density regions into low-density regions, and vice versa. That is just as good a set of initial conditions as the first one!

Suppose we had initial conditions that had a meatball topology. The high-density regions would be isolated meatballs and the low-density regions would be one connected soup. If we multiplied all our initial fluctuations by −1, it would turn the high-density meatballs into low-density voids (holes) and the soup into one connected, high-density piece of cheese. If random fluctuations *always* made a meatball topology, then multiplying these fluctuations by −1 would turn them into equally random fluctuations but suddenly with a Swiss cheese topology—that's inconsistent. Thus, random fluctuations cannot produce an initial meatball topology. Random fluctuations cannot produce an initial Swiss cheese topology either—multiplying those fluctuations by −1 would turn them into a meatball topology. From this reasoning, neither meatballs (the American school of cosmology) nor Swiss cheese (the Russian school) could be produced by the random initial conditions predicted by inflation! Both led to a logical contradiction. By contrast, a sponge can have insides and outsides that are the same. Imagine our marine sponge example again, with water percolating through it. Multiply the fluctuations by −1, and geometry of the new high-density regions (concrete) is just as spongelike as before. Random initial conditions can produce this. Thus, a spongelike geometry with insides and outsides that look alike is what is required for Gaussian random-phase initial conditions. I got this idea from my high school science project.

A High School Science Project

I was always drawn to geometrical problems. After eighth grade, I attended a summer program in mathematics at Rollins College in Florida sponsored by the National Science Foundation and run by Professor Bruce Wavell. This program had wonderful courses on mathematical logic and a course on special relativity based on Max Born's book. I frequented the library and found exciting books on four-dimensional geometry. At this summer program I began a project on the geometrical arrangement of atoms in metallic crystals.

In a metal there is a periodic array of atoms, but since all the atoms are identical, one might expect the arrangement of atoms around any given atom to be the same. I thought such periodic structures could be related to volume-filling packings of identical polyhedrons. As we have discussed, we can stack cubical boxes to fill a warehouse leaving no empty spaces. Imagine an atom fitting inside each box. The nucleus of each atom would be at the center of each box. The points within a given box represent the set of points in space closer to the center of that atom than to the center of any other atom. A warehouse filled with such cubical boxes with one atom in each would produce a *cubic* crystal structure. The atoms would be located at the vertices of a cubical jungle gym. But what other kinds of identical cells could equally well fill an infinite warehouse, leaving no gaps? I found a thick book showing all kinds of polyhedrons and set out to test each one. Using spherical trigonometry, I wrote a computer program to figure out which polyhedrons would fit together to fill space. I ran this program on an early IBM 1620 computer, which used punched cards and was about the size of a washer and dryer. An iPhone would exceed its computing power by a large factor today, but it was the state-of-the-art computer available for general use at that time.

I found a number of different polyhedrons that could stack together to fill space. For example, triangular prisms would fit together in this way: a triangular prism has a triangular top and bottom with three square sides connecting them. Lay out equilateral triangles to tile a plane; identical triangular prism boxes constructed on top of each triangle make a layer that covers the entire floor of the warehouse; then put a second layer on top of that and continue until the warehouse is filled. This makes a pattern like the arrangement of carbon atoms in graphite.

One particularly interesting solution used *truncated octahedrons* to fill space. A truncated octahedron is an octahedron with its corners cut off—it has six square faces and eight hexagonal faces.

If I have a warehouse filled with truncated octahedral boxes, as depicted in Figure 6.1, the atoms centered in each box will form a *body-centered cubic array*. It's as if you had a jungle gym with atoms at all

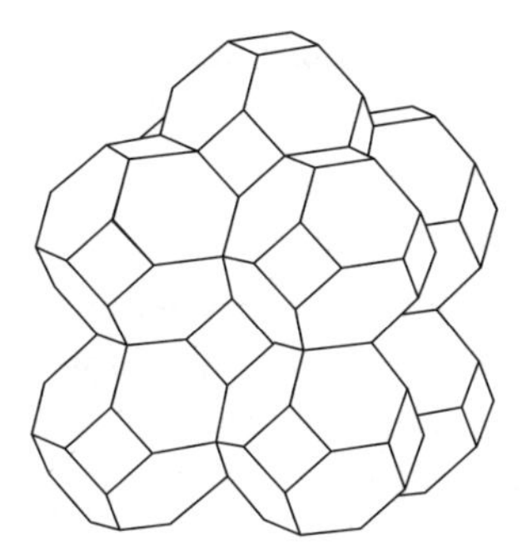

Figure 6.1. How truncated octahedrons can be stacked to fill space in a body-centered cubic pattern. Eight truncated octahedrons form a cube—with another one filling the center of the cube. One more truncated octahedron sits on top of that central one, thereby starting the next layer. This pattern can be extended indefinitely to fill space. Note that the vertex on the left, between the two frontward-facing squares on the left, is surrounded by four hexagons in a butterfly pattern. These four hexagons form a saddle-shaped surface around the vertex. This will prove to be important later. (Credit: A.J.S. Hamilton, J. Richard Gott, and D. Weinberg, *Astrophysical Journal*, 309: 1, 1986)

the intersection points between rods and *also* an atom suspended in the middle of each cubic open space in the grid. This is the structure of metallic sodium. I noted, in addition, that the shape of the occupied electron orbitals in metallic sodium actually looks approximately like a truncated octahedron.

This project won first place in Physical Sciences at the Kentucky State Science Fair and that got me a place in the National Science Fair International, which had entrants from all over the United States and from a number of other countries. There, my crystal project won a first prize from the American Metallurgical Society and a runner-up award for a trip to Japan from the U.S. Navy to exhibit at the Japanese Science Fair. Over the summer I improved my project by doing X-ray diffraction studies of metallic sodium to prove that it indeed had a body-centered cubic pattern and submitted it to the Westinghouse Science Talent Search.

The Westinghouse Science Talent Search (now the Intel Science Talent Search) has been nicknamed the "Nobel Prize" of high school science competitions. From 300 semifinalists, 40 winners are brought to Washington, D.C., where they are interviewed over several days by a panel of judges. I was selected as one of the 40 winners. One of the other winners my year was Ray Kurzweil—well known today for his invention of the Kurzweil reader for the blind and for his ideas on artificial intelligence and the implications of increasingly powerful computer technology for the future.

One of our judges was Glenn Seaborg (who discovered plutonium and other transuranic elements). He seemed to like my project and asked me questions about sodium, which I was able to answer. Finally came the awards banquet, a black-tie affair held in a fancy hotel ballroom. Ten scholarship winners were read off one by one, starting with tenth place. In the end, I came in second. I was thrilled—with a few names called in reverse order, you are happy to hear your name come up at all![1]

The scholarship from the contest paid for about half of my Harvard undergraduate education. But, in a way, the most important thing occurred before I got to Washington, when I was rushing to prepare my exhibit for the judges. I had made a plastic model of transparent truncated octahedrons stacked to fill space (like Figure 6.1). Inside each was a Styrofoam ball representing a sodium atom. The truncated octahedrons themselves were made from Plexiglas® sheets that I had cut into hexagons and squares. They had beveled edges that I had cut on a bench saw at the proper angle so that they would glue together to form the truncated octahedrons. I had assembled a set of hexagons glued together in pairs, and I noticed that if I put two pairs of beveled hexagons together, the four hexagons formed a *saddle-shaped* surface. I knew that a saddle shape corresponded to a negatively curved surface. (A sphere has positive curvature, a plane has zero curvature, and a Western saddle has a negative curvature.) This fact would later have an important application for large-scale structure because sponges turn out to have negatively curved surfaces.

A triangle drawn on a Western saddle will have a sum of angles that is less than 180°. The circumference of a circle drawn on a Western saddle will be *larger* than 2π times its radius. That's because the saddle goes up and down as you circle the center, making that circumference longer than it would be on a flat plane.

Mathematician Carl Friedrich Gauss proved that the intrinsic curvature of a 2D surface is equal to $1/r_1r_2$, where r_1 and r_2 are the principal radii of curvature of the surface.

For example, a flat plane has infinite radii of curvature, so $1/r_1r_2 = 0$, giving a flat plane zero curvature. A cylinder also has a curvature given by $1/r_1r_2$, but in this case, r_1 is the radius of the cylinder, while r_2, the radius of curvature along its length, is infinite, because the cylinder is straight along the direction of its length. No matter the value of r_1, r_1 times infinity is infinity. And 1 over infinity is 0, so the Gaussian curvature of a cylinder is also 0. You can make a cylinder out of a flat piece of paper without distortion.

On a sphere, the two principle radii of curvature, r_1 and r_2, are both equal to the radius of the sphere, and both point in the same direction. If you sit on top of a sphere, your two legs can hang down on opposite sides, while the sphere also curves down in the front and back direction as well. It curves downward in both perpendicular directions. The curvature of a spherical surface is thus positive, equal to $1/r^2$, where r is the radius of the sphere.

In a Western saddle, although the saddle curves downward from side to side where your legs go, it curves upward from front to back to fit on the horse's back. Thus the two radii of curvature in a Western saddle point in opposite directions and their product is negative, giving the Western saddle a negative curvature.

So when I noticed that those four hexagons glued together (where two of the truncated octahedrons were joined together) formed a saddle-shaped surface, I immediately related it to a negatively curved surface (see Figure 6.1).

Positively curved surfaces like the sphere can be approximated by regular polyhedrons. The cube is a very rough approximation to a sphere. A cube is made up of squares meeting 3 at a point. In a cubical room, the ceiling meets 2 side walls at a vertex—look up at the corner of the room you are sitting in to verify this. Each square face of a cube has a 90° angle at its corner. So at the corner of a cube, three square faces meet at a vertex, and each has a 90° angle at its corner, making the sum of the angles around a point at the corner 3 × 90°, or 270°—that is 90° *less* than one would get in a plane. A plane can be tiled by squares 4 around a point to make a checkerboard pattern, where the sum of angles around each point is therefore 4 × 90°, or 360°—just what one would expect. On a polyhedron that approximates a positively curved surface, the sum of angles around a vertex is less than 360°. The cube has an *angle deficit* of 90° at each vertex. What about my 4 hexagons? Hexagons have 120° angles at their vertices, which makes the sum of angles around a vertex in my saddle-shaped surface 4 times 120°, or 480°! That is more than 360°. There is an *angle excess* of 120° at the vertex. It all made sense. If one has polygons approximating positively curved surfaces (as occurs in the 5 regular polyhedrons), they will have less than 360° around a point; if they approximate a plane surface (as in a checkerboard), they will have exactly 360° around a point; and if they approximate a negatively curved surface, they will have more than 360° around a point. (In fact, angle deficits and angle excesses would later enable us to write a computer program to measure the curvature of the complex surfaces we encounter in studying large-scale structure in the universe. Unknowingly, I was learning in high school the tools I would need later to study the topology of large-scale structure.)

Could this pattern of four hexagons around a point be extended to continue the negatively curved surface? I noticed that it could! In fact, if you just took my plastic model of truncated octahedrons filling space and *deleted* all the squares, you would be left with a single convoluted surface—composed only of hexagons meeting four at a point in every case. The hexagons formed *a spongelike surface* that divided space into two equal parts that were identical to each other but interlocked. In high school I already realized that this surface made of hexagons had numerous holes, like donut holes. I had found a new regular polyhedron—an infinite spongelike one—composed of hexagons meeting four at a point. The astronomical connection is this: it is such a spongelike surface that inflation would require for the geometry of the median-density contour surface separating the high- and low-density regions.

Regular polyhedrons are defined as having regular polygons (i.e., closed planar figures having equal sides meeting at equal angles) as faces, with the same configuration of faces around each vertex. Finding a new one was exciting because only five regular polyhedrons had been known from ancient Greek times (Figure 6.2). Each has vertex angles totaling less than 360°. These were the only possibilities that added up to less than 360° to make finite closed figures roughly approximating positively curved spheres.

But, in addition, there were three *regular planar networks* (shown in Figure 6.3). Johannes Kepler recognized that these long-known planar networks were like regular polyhedrons but with an infinite number of faces.

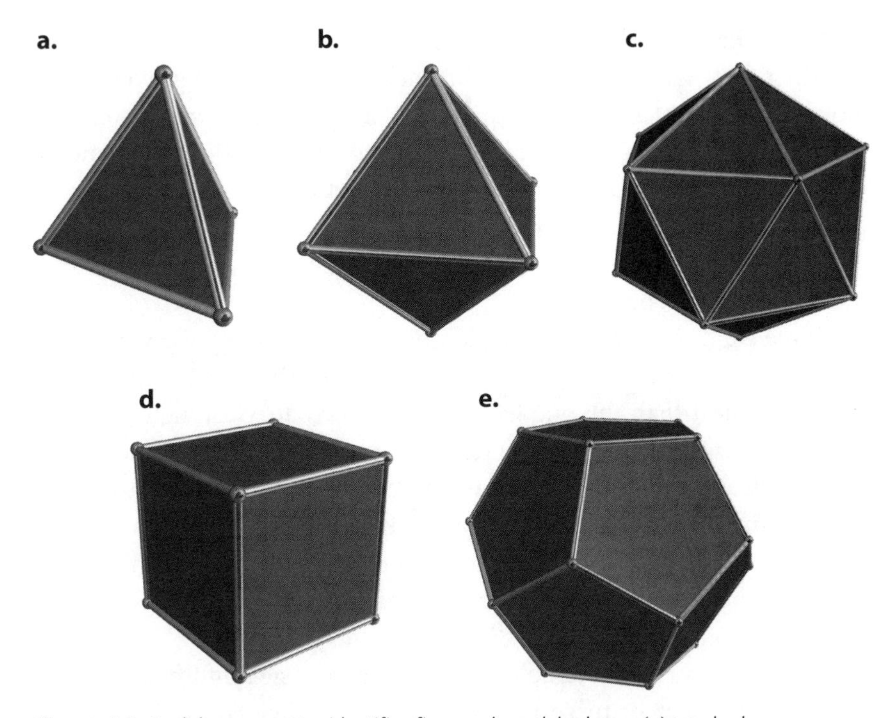

Figure 6.2. Euclidean geometry identifies five regular polyhedrons: (a) tetrahedron, (b) octahedron, (c) icosahedron, (d) cube, and (e) dodecahedron. These are polygon networks with (a) triangles, 3 around a point; (b) triangles, 4 around a point; (c) triangles, 5 around a point; (d) squares, 3 around a point; and (e) pentagons, 3 around a point, respectively. (Credit: Robert Webb's *Stella* software is the creator of these images: http://www.software3d.com/Stella.php [here adapted as black and white])

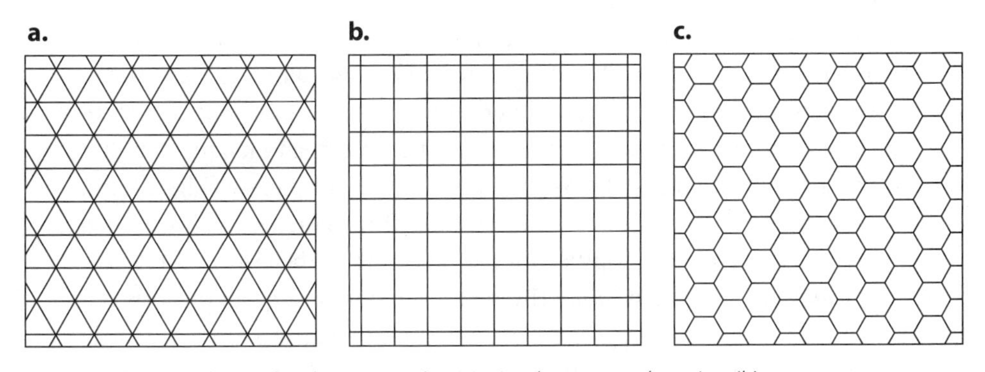

Figure 6.3. The regular planar networks: (a) triangles, 6 around a point; (b) squares, 4 around a point; and (c) hexagons, 3 around a point. (Adapted from: R. A. Nonenmacher)

Now I had found a regular polygon network (hexagons, 4 around a point) that was negatively curved (having a sum of angles around a point of 480°, which is greater than 360°) and was also spongelike, with an infinite number of faces and an infinite number of holes. I quickly found six other spongelike networks. I called these *pseudopolyhedrons* (Figure 6.4), after the pseudosphere which is a surface of constant negative curvature encountered in the non-Euclidean geometry of Nikolai Lobachevsky and Janos Bolyai.

These were seven regular polygon networks that were every bit as legitimate as the ones known from ancient times! This was a much better science project than the one I was going to take to Washington for the Westinghouse contest. I was just turning eighteen at the time. I wondered if I should substitute this project for the exhibit I was planning to take. I decided there was not enough time to make it. But I did think that I might turn this into an exhibit for the science fair later in the spring. This meant constructing cardboard models of all these pseudopolyhedrons. I made the convoluted surfaces white on one side and red on the other. (In Figure 6.4 these surfaces are colored white on one side and black on the other.) A number of these pseudopolyhedrons divided space into two equal parts—these were *sponges* whose insides and outsides were identical: squares, 6 around a point; hexagons, 4 around a point; hexagons, 6 around a point; pentagons, 5 around a point; and triangles, 10 around a point.

In the end, my pseudopolyhedron project won first place in mathematics in the Louisville Science Fair and the top prize over all, so I got to go to the National Science Fair International held in St. Louis. My wonderful math teacher Mrs. Ruth Pardon got to go along as well. There, it won first place in mathematics, and one of three grand prizes, a trip to Japan sponsored by the U.S. Navy. I ended up traveling to Japan after all! I took the trouble to translate my project's labels into Japanese (Figure 6.5).

After I got to Harvard, math instructor Tom Banchoff [2] encouraged me to submit my paper on pseudopolyhedrons to the *American Mathematical Monthly*, which I did. The referee's report was quite positive but noted that three of my polygon networks had been discovered before! The reference was to a paper, which I had never heard of, by H.S.M. Coxeter in 1937. That paper described how the first of these figures to be discovered—squares, 6 around a point—was found in 1926 by John

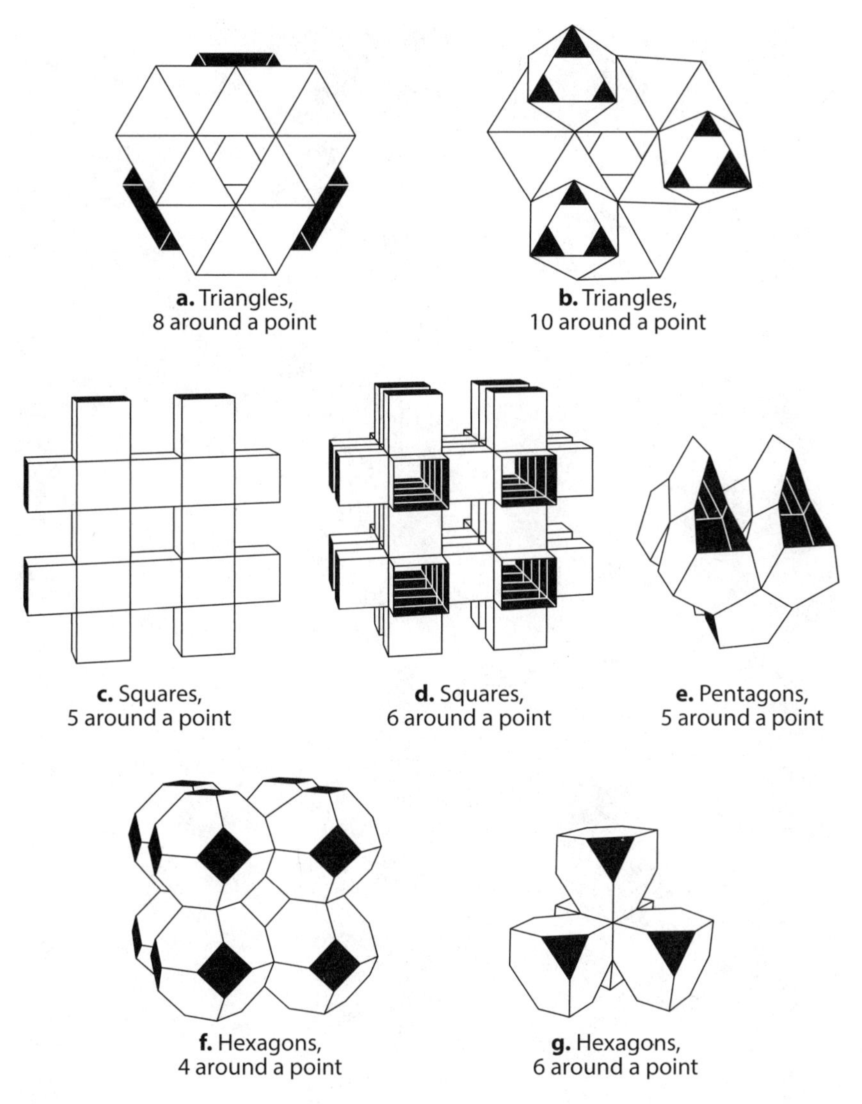

Figure 6.4. Regular pseudopolyhedrons expand the class of regular polyhedrons. (Credit: J. Richard Gott, *American Mathematical Monthly*, 74: 497, 1967)

Petrie, who also discovered hexagons, 4 around a point (the one I found first). Therefore, Petrie gets credit for discovering this entire class of figures. Coxeter himself discovered hexagons, 6 around a point. They did this work in 1926 when they were both 19. In addition to demanding that the configuration of polygons around each vertex be identical (as I

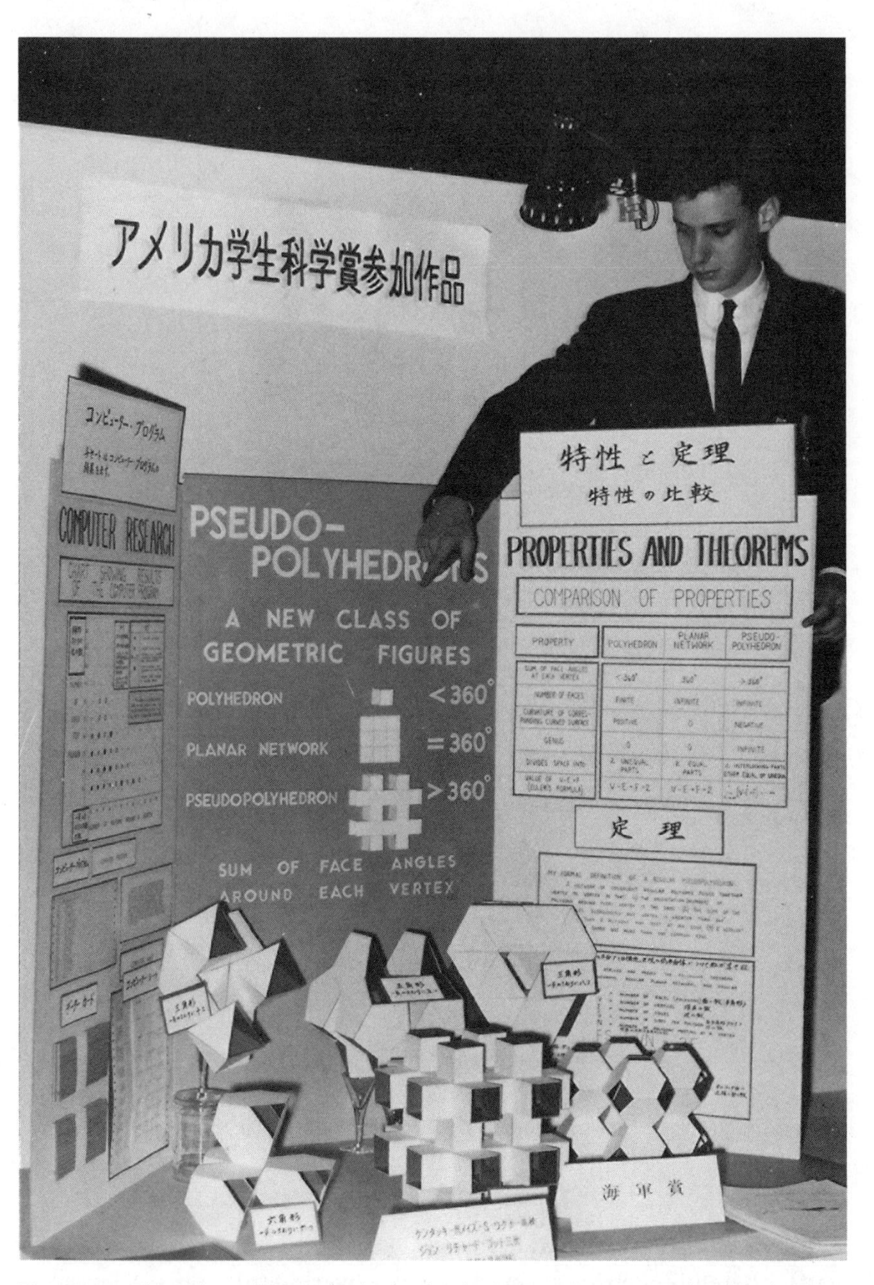

Figure 6.5. The author exhibiting his high school science project on pseudopolyhedrons at the 1965 Japanese Science Fair. (Credit: J. Richard Gott, personal collection)

did), their criteria for regularity also demanded that the angles between all adjacent pairs of faces be equal. With those conditions they were able to prove that the three examples they found were the only regular figures of this type. They called them *regular skew polyhedrons*. I was happy to add the Coxeter-Petrie reference. My paper was still publishable, the referee said, because I had discovered four new pseudopolyhedrons. I still required the configuration of polygons around each vertex to be identical but allowed the angles between adjacent faces to vary. I rediscovered all three structures discovered by Petrie and Coxeter as well as finding four new ones allowed by my more lenient rules. My paper appeared in print in 1967. It was my first published scientific paper. When Siobhan Roberts wrote her definitive biography of Coxeter, *King of Infinite Space*, in 2006, I was happy to contribute my story of the astronomical applications these figures later had in understanding the distribution of galaxies in space.

Additional regular pseudopolyhedrons have been discovered by the noted crystallographer A. F. Wells: including triangles, 7 around a point; triangles, 9 around a point; and triangles, 12 around a point. These are illustrated in Wells's 1969 paper and his 1977 book, *Three Dimensional Nets and Polyhedra*. He, like I, did not demand equal angles between adjacent faces. All are spongelike with an infinite number of faces and an infinite number of holes.

Melinda Green rediscovered my pentagons, five around a point, and has illustrated many pseudopolyhedrons on her geometry Web page. Her illustration of pentagons, 5 around a point, appears in Color Plate 1.

Avraham Wachmann, Michael Burt, and Menachem Kleinman have discovered many semiregular spongelike polyhedrons, composed of polygons of more than one kind, for example, two squares and two hexagons around each point. These are illustrated in their 1974 book *Infinite Polyhedrons*.

In mathematics, the *Schwarz P surface* is a negatively curved surface that approximates the geometry of the pseudopolyhedron having hexagons, four around a point; it divides space into two equal spongelike parts. In biology, the Schwarz P surface has been observed in the biological membranes of various cells by transmission electron microscopy. Thus, microscopic spongelike surfaces have been observed in nature.

A Spongelike Universe

Would the universe at large scales also be spongelike? That is what inflation seemed to predict. If random quantum fluctuations produced the small fluctuations present in the initial conditions, then the geometry of those regions above average and below average in density had to be identical. The geometry of a sponge could achieve this, as illustrated by the pseudopolyhedron squares, six around a point (Figure 6.6). The high-density parts would constitute one connected sponge with many holes, and the low-density regions would form a complementary sponge of low-density voids connected by low-density tunnels.

I set out to test this. I called up Adrian Melott at the University of Kansas, who had done some of the world's largest *N*-body simulations of growth of structure in the universe. I knew from his papers that he had programs for drawing density contour surfaces in 3D and that these could be applied both to his initial conditions and to the outcome today.

When measuring the topology of large-scale structure today, one has the positions of individual galaxies to work with—a set of points in 3D space. We are interested in how these points are clustered: where are the regions where there are more points than average and where are there fewer than average? So we must take averages. We do this by smoothing, or blurring, the data to produce a smoothly varying average density field. It is important to smooth the data over a length scale that is at least as large as the average separation between galaxies and on a scale where

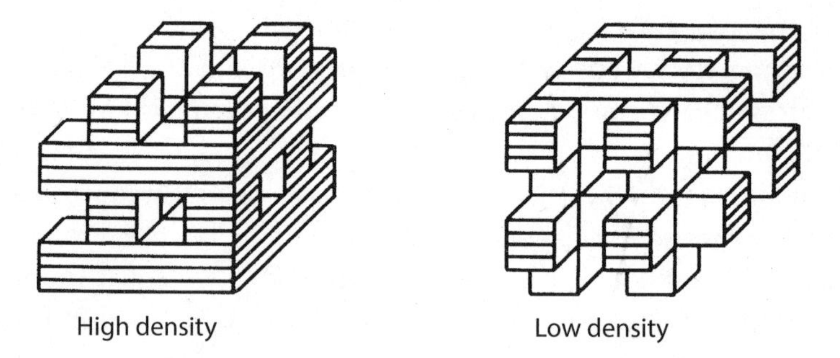

Figure 6.6. This pseudopolyhedron (squares, 6 around a point) divides space into equivalent, spongelike high- and low-density halves. (Credit: J. Richard Gott, A. L. Melott, and M. Dickinson, *Astrophysical Journal*, 306: 341, 1986)

the average density fluctuations from place to place are small compared with the average density for the universe as a whole. In practice, this means smoothing, or blurring, the data over a scale of at least 24 million light-years. It's like producing a population density map of the United States by averaging over counties rather than showing the points representing each of the people.

First, we smoothed the random initial conditions of Melott's simulation based on inflation and plotted the median density contour. Half the volume was on the high-density side of this contour and half was on the low-density side. This contour surface was spongelike, just as I had predicted it might be! The high-density side was multiply connected with many holes, and the low-density side was also multiply connected and *interlocking* with the high-density side. Space was divided into two complementary sponges—like the marine sponge and the water. This was, of course, a random sponge; my pseudopolyhedrons were periodic sponges. But both were sponges that divided space into two complementary parts.

Then we watched the structure grow in the simulation and waited until the simulation reached the present day. The initial fluctuations were so small as to be barely visible, but gravity worked on them to make them bigger, with the high-density parts getting ever denser relative to the average density, while the low-density parts emptied out. Eventually the very densest parts in the initial conditions stopped expanding with the universe and collapsed to form clusters of galaxies, while low-density voids also appeared. In the end, the clusters were connected by filaments of galaxies to make a spongelike pattern—what we now call the *cosmic web*. The low-density region cleared out to make a complementary low-density sponge—empty voids connected by low-density tunnels. We smoothed the map showing the density at the present epoch by blurring it on a scale of 94 million light-years. Then we had the computer again draw the median density contour. This contour contained half the volume on the high-density side and half the volume on the low-density side, just like the one we computed in the initial conditions. It was also a sponge. Furthermore, it was almost exactly the *same* sponge we started with—it had the same holes in the same places (Figure 6.7). They had moved a little and their shapes had changed a little, but their topology looked the same. One could recognize the same pattern of holes, implying that the

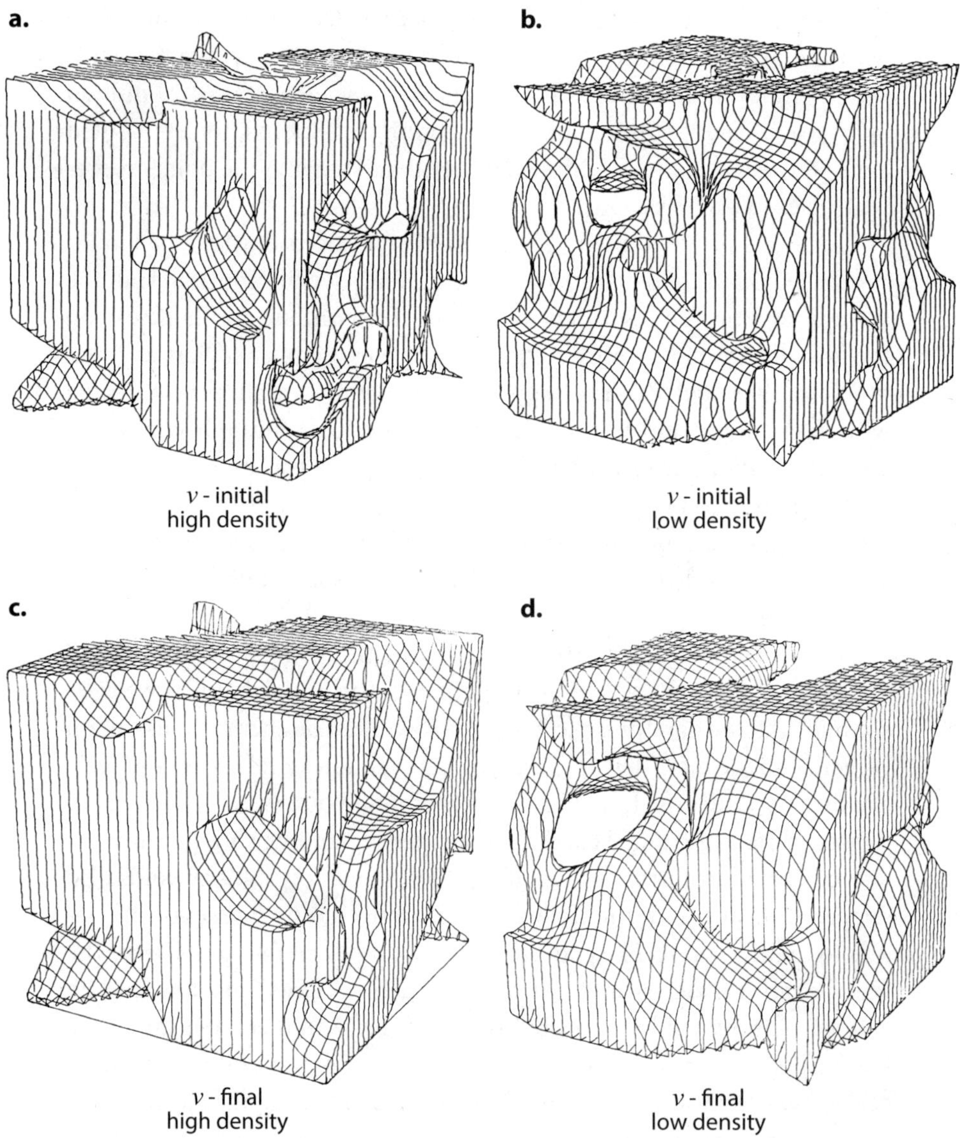

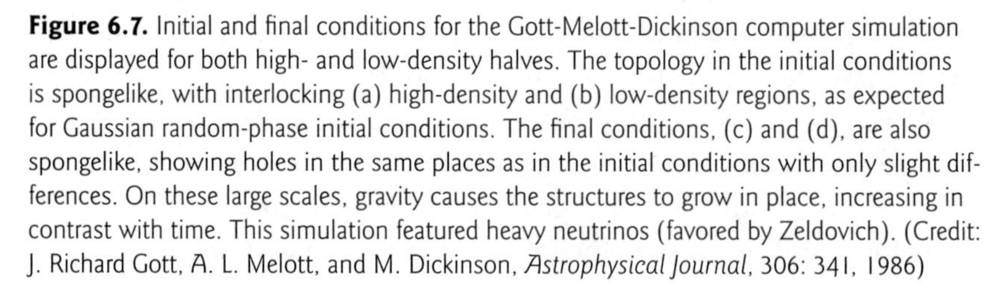

Figure 6.7. Initial and final conditions for the Gott-Melott-Dickinson computer simulation are displayed for both high- and low-density halves. The topology in the initial conditions is spongelike, with interlocking (a) high-density and (b) low-density regions, as expected for Gaussian random-phase initial conditions. The final conditions, (c) and (d), are also spongelike, showing holes in the same places as in the initial conditions with only slight differences. On these large scales, gravity causes the structures to grow in place, increasing in contrast with time. This simulation featured heavy neutrinos (favored by Zeldovich). (Credit: J. Richard Gott, A. L. Melott, and M. Dickinson, *Astrophysical Journal*, 306: 341, 1986)

universe "remembered" what its initial conditions looked like (see Figure 6.7). The important point here is that the topology on large scales is preserved, despite the strong evolution of clustering on small scales. Smoothing more or less recovers the initial conditions. In the computer simulation, those initial conditions were the random fluctuations predicted by inflation. With this topology tool, we could now look at the universe today to deduce what it looked like at the beginning—to check whether it had the spongelike initial conditions expected from inflation.

The next step was for me to enlist my then-undergraduate student Mark Dickinson (now at Kitt Peak National Observatory in Arizona) to look at the observational data and apply the same test. Using data from the Center for Astrophysics (CfA) survey, we constructed a uniform cubic sample stretching beyond the local Virgo Supercluster of galaxies. including all galaxies above a certain luminosity all the way out to the sample's outer edge. We thus had a complete sample of bright galaxies inside the cubical region we were exploring, with our Milky Way Galaxy located at the bottom-front corner of the cube. We could get a clear view looking out of the plane of our galaxy into a cubic survey region beyond. Mark then smoothed the data on a scale of 47 million light-years and had the computer draw the median density contour (see Figure 6.8). It was spongelike—just like the simulations.[3] The high-density parts, which included the Milky Way and the Virgo Supercluster, were on the high-density side of the spongelike contour, and the low-density parts were on the other side. It was a very small sample, but we felt it was a eureka moment nevertheless, and we knew that it was only a matter of time before much-larger observational samples would become available, allowing us to test our theory in great detail. Our paper (Gott, Melott, and Dickinson 1986) was titled "The Sponge-like Topology of Large-scale Structure in the Universe" and was published in the *Astrophysical Journal.*

We wanted to quantify the *topology* of these structures. Topology is the field of mathematics studying those properties of geometric figures that remain unchanged under distortion. Topologists talk about *genus.* A donut has a genus of 1. It has 1 hole. A coffee cup with one handle also has a genus of 1. To a topologist, a donut and a coffee cup look just alike! One can be distorted into the other without breaking. If you had a donut made out of clay, you could continue molding it without ever breaking

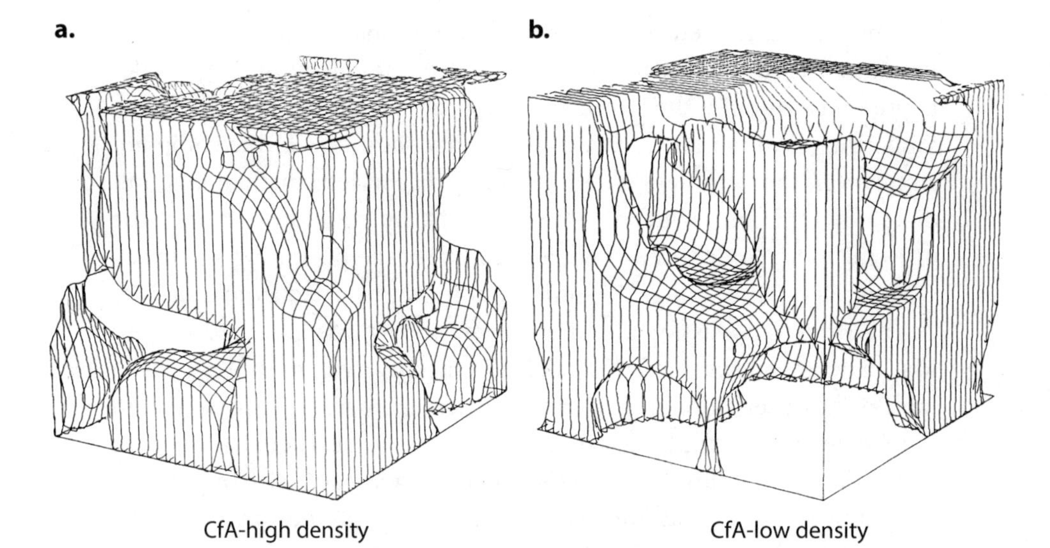

Figure 6.8. Observations (from the CfA sample) are shown divided into (a) high-density and (b) low-density halves. We studied a cubical observational sample, 140 million light-years on a side, with an effective smoothing scale of 47 million light-years. Earth is at the bottom front corner. The topology is spongelike. (Credit: J. Richard Gott, A. L. Melott, and M. Dickinson, *Astrophysical Journal*, 306: 341, 1986)

it until you had made a coffee cup with a handle. But if you cut through the coffee cup handle, its genus becomes 0, because the cup now has no holes. A sphere has a genus of 0. You can take a sphere of clay and mold it into a coffee cup (without a handle). The cup has a depression where the coffee goes, but that is not a hole—it doesn't go all the way through. Eyeglass frames have 2 holes: they have a genus of 2. A trophy with two handles has a genus of 2 as well. If you cut through one of its handles, you destroy that hole and reduce the genus by 1, leaving it with a genus of 1, like a coffee cup. So genus corresponds to the number of holes in an isolated figure; alternatively, one can say it is equal to the maximum number of complete cuts you can possibly make in the figure without it falling apart into two pieces.

For cosmology we needed a new definition of genus—one that could be applied to many structures, not just one. We *redefined* the genus of a density contour surface as

Santa Clara County Library District
Checkout Receipt
Saratoga Library
08/14/17 05:45PM
24/7 Renewals:
Online at www.sccl.org
Telecirc: 800-471-0991
PATRON: 32788
The cosmic web : mysterious architecture
33305234720833 Due: 09/05/17
TOTAL: 1
Thank you for visiting the library!

genus = number of donut holes − number of isolated regions.

By this *new* definition, a donut has a genus of 0, because it has one donut hole and is itself one isolated region (1 − 1 = 0). A sphere (e.g., a spherical density contour surrounding an isolated cluster of galaxies) would have a genus of −1 under this new definition (0 holes − 1 isolated region = −1). Two spheres would have a genus of −2 because they have no donut holes and would enclose two isolated regions (0 − 2 = −2). If you cut a sphere in half, you would lower its genus by 1: as a single sphere, it was 1 isolated region with no holes and a genus of −1; after it was cut, it would be two isolated hemispheres (0 holes − 2 isolated regions), giving a genus of −2. Every time you cut, you lower the genus by 1, just as before. With this new definition of genus, we can keep track of how many holes are found in the density contour surface as well as how many isolated regions it is broken up into. If we had a meatball topology, for example, the median density contour surface would show isolated clusters and we would have many isolated regions and a negative genus. If we had a Swiss cheese topology, the median density contour would likewise have many isolated voids and a negative genus. We could distinguish these two topologies by noting whether the isolated regions contained the high- or low-density parts. But if we had a spongelike topology, the median density surface would be one convoluted piece, multiply connected, with many holes. It would have *no isolated regions but many holes*, making its genus positive. We were able to prove (using a famous Gauss-Bonnet theorem from mathematics) that the newly defined genus could be calculated by integrating the curvature over the contour surface and dividing by -4π. To understand this better, let's consider some properties of curvature.

Measuring Curvature—David Weinberg's CONTOUR3D Program

A sphere has a uniform curvature of $1/r^2$ over its entire surface. The area of the sphere is $4\pi r^2$, so if we integrate, or add up, the curvature over the entire area of the sphere, we get $1/r^2$ times $4\pi r^2$, or 4π. This is independent of the radius of the sphere. If the sphere is twice as big, its curvature is ¼ as large but its area is 4 times as big, so the answer is 4π, just as before. If we divide 4π by -4π, we will get a genus of -1 for the sphere (one isolated region with no holes). It doesn't depend on how big the sphere is.

For a sphere the curvature is uniformly spread over the entire surface. But for a polyhedron like a cube, all the Gaussian curvature is concentrated in the vertices. The faces of the polyhedron are flat, and the edges are like bent cylinders, which also have zero curvature as defined by Gauss. If you have ever tried to flatten some cardboard boxes for recycling, you will realize that the problem occurs not at the edges but at the vertices—you have to do some tearing to get the job done there. In a polyhedron, the integral of the curvature over a vertex is equal to the *angle deficit* at that vertex. Mathematicians measure angles in radians, where 2π radians are taken to be equal to 360°. The angle deficit at each vertex of the cube is 90°, or $\pi/2$ radians. The cube has 8 vertices, 4 on the top and 4 on the bottom, so the total angle deficit for the cube is 8 times $\pi/2$ radians, or 4π radians. Thus the integral of the curvature over the entire cube is 4π, just as it was for the sphere. The cube is one isolated region with no holes, so by our definition it should have a genus of -1, just like the sphere. To a topologist, a sphere and a cube look the same.

My spongelike pseudopolyhedrons, by contrast, have angle *excesses* (i.e., more than 360°) around each vertex. An angle excess is the opposite of an angle deficit—in other words, a *negative* angle deficit. And this corresponds to a negative curvature. When we integrate the curvature over our spongelike pseudopolyhedrons, we will get a negative number that will tell us how many holes we are looking at. In our paper, we explained how to write a computer program that would do this calculation automatically. The volume of the computer simulation is divided into little cubic *voxels*, or 3D volume elements, like pixels but in 3D. The boundary between the high-density and low-density voxels will be a corrugated surface of square faces. Our program would then look at each vertex in this corrugated surface: if three squares came together at a point, it would have a positive curvature like the cube; if 4 squares came together at a point, it would have zero curvature like a plane; and if 5 or 6 squares came together at a point (as they do in the pseudopolyhedrons squares, 5 around a point, or squares, 6 around a point), then the curvature would be negative. Polygon networks (as in my high school science project) were now going to be used to measure the curvature. The program just adds up the contributions from each vertex to determine the answer. Dividing the answer by -4π would tell us the number of holes minus the number of isolated regions the density contour surface had. My graduate student at the time, David Weinberg (now a professor at Ohio State University), wrote this program, which we called CONTOUR3D. It allowed us to calculate the genus of any contour surface.

The Genus Predicted by Inflation

I next enlisted Andrew Hamilton at Princeton (now at the University of Colorado) to help with calculating what the genus predicted by inflation would be. I designed the voxels for this calculation—in this case, truncated octahedrons (as it happens, just like the cells for sodium atoms in my high school Westinghouse project)—and he did the calculation. Given the probabilities of different voxels being above or below the contour threshold, one could calculate the probability of a particular curvature occurring at each vertex and, therefore, the genus. Inflation predicted *Gaussian random-phase fluctuations* in the initial conditions because these were produced by random quantum effects. Think again of random waves in an ocean. Inflation also told you how much power to expect at different wavelengths. The power is proportional to the square of the height of a typical wave of a certain wavelength. Hamilton's calculations gave a formula for the genus. If you want to see what it looks like, here it is: genus = $A(1 - \nu^2)\exp(-\nu^2/2)$. It is a simple formula. A is a positive constant that depends on the amount of power at different wavelengths, ν is the number of standard deviations above the mean for the contour surface, and $\exp(x)$ (the number $e = 2.71828\ldots$ raised to the power x) is the exponential function, which is always positive. The function $\exp(-\nu^2/2)$ is a bell-shaped or Gaussian distribution, which is always positive. Suffice it to say, at the median density contour surface ($\nu = 0$), which divides space into two equal parts, the genus was *always* positive—indicating a spongelike topology. If, however, we look at the highest-density regions, comprising only 7% of the volume ($\nu = 1.45$), we will see a negative genus—only isolated clusters. If we look at the lowest-density regions, comprising only 7% of the volume ($\nu = -1.45$), we will find isolated voids—also a negative genus. This gives a symmetric genus curve, as we go from low-density contours toward high-density contours: negative genus (isolated voids) at first, positive genus (in the middle-density ranges), and then negative genus again (isolated clusters)—creating a W-shaped curve. Reverse the sign of initial conditions, replacing high-density regions with low-density regions, and the curve stays the same—as must be true for random quantum fluctuations. Now we had a whole theoretical curve to test against both our simulations and against actual observational data sets.

After completing this calculation, Hamilton found that Doroshkevich in 1970 had derived essentially the same formula by an independent method. Doroshkevich even remarked in passing that the contour surface for $\nu^2 < 1$ would be complex and multiply connected, but he was just interested in galaxy formation and in using the formula to count peaks for $\nu > 1$. Bardeen, Bond, Kaiser, and Szalay in 1986 also used it to count high-density peaks, but no one had applied it to examine the median density contour in galaxy clustering. It turns out that mathematician R. J. Adler had calculated these topology measures in N dimensions in 1976 and discussed them in his 1981 book, *The Geometry of Random Fields*. They also go by the name of *Minkowski functionals*. Hikage and colleagues (2003, 2006), who have measured topology following our work, have used this nomenclature for the genus, as did Mecke and colleagues (1994). In 3D, the genus, as we are defining it, for a density contour surface is related to the *Euler characteristic* $(V - E + F)$. A 3D density contour surface, if pixelated to make a network of polygons, will have a number of faces (F), edges (E), and vertices (V) satisfying the relation:

$$\textit{genus} = -(V - E + F)/2 = -\text{Euler characteristic}/2,$$

where the 3D genus is as *we* have defined it (number of holes – number of isolated regions).

That the value of $(V - E + F)$ was related to topology was proven by the great mathematician Leonhard Euler (1707–83). For example, a cube has 8 vertices, 12 edges, and 6 faces, giving it an Euler characteristic of 2 and a genus of -1 for one isolated region, by our definition. In my high school science project, using ratios of vertices, edges, and faces, I had already realized that the Euler characteristic of my infinite spongelike polyhedrons was negative and infinite, allowing me to show that they had an infinite number of holes.[4] Euler's formula is a piece of genius.

Thus, we have brought some famous old mathematics to bear on testing whether the universe started with Gaussian random-phase initial conditions as predicted by inflation.

Measuring the Genus Topology

We calculated the genus at various density contours in the initial conditions of our computer simulation and found excellent agreement with the formula. Since the initial conditions were calculated according to the initial conditions for inflation, that was no surprise. But the results also matched the formula with little difference *at the present epoch*. If an appropriately large smoothing length was applied, the density contours of cold dark matter in terms of volume fraction at present still looked quite like the similarly smoothed density contour surfaces in the initial conditions. The contrast in the picture at the present epoch was larger, but the topology of the map was much the same as in the initial conditions (as shown in Figure 6.7). It's as if the mountains got higher and the valleys got lower while staying in the same places. Gravity grows the fluctuations in place. In the cold dark matter model, we expect galaxies to track the cold dark matter, but we expect them to be *biased* toward forming in the highest density dark matter regions. As long as the probability of making a galaxy was a monotonic (ever-increasing) function of the density of the cold dark matter, the topology of the biased (galaxy) map and the cold dark matter map should be similar. The voxels of cold dark matter and galaxies should be ranked similarly, and the median density contour by volume should be similar. Thus, great clusters today are located where tiny density enhancements in the cold dark matter occurred in the initial conditions, and great voids today are located where tiny density decrements in cold dark matter occurred in the initial conditions. Figure 6.9 shows the results we obtained studying a computer simulation that assumed cold dark matter. The spongelike median density contour (50% low—50% high) that divides space into two equal volumes hardly moves at all between the initial conditions and between the final and final biased distributions. The holes are all in the same places.

a.

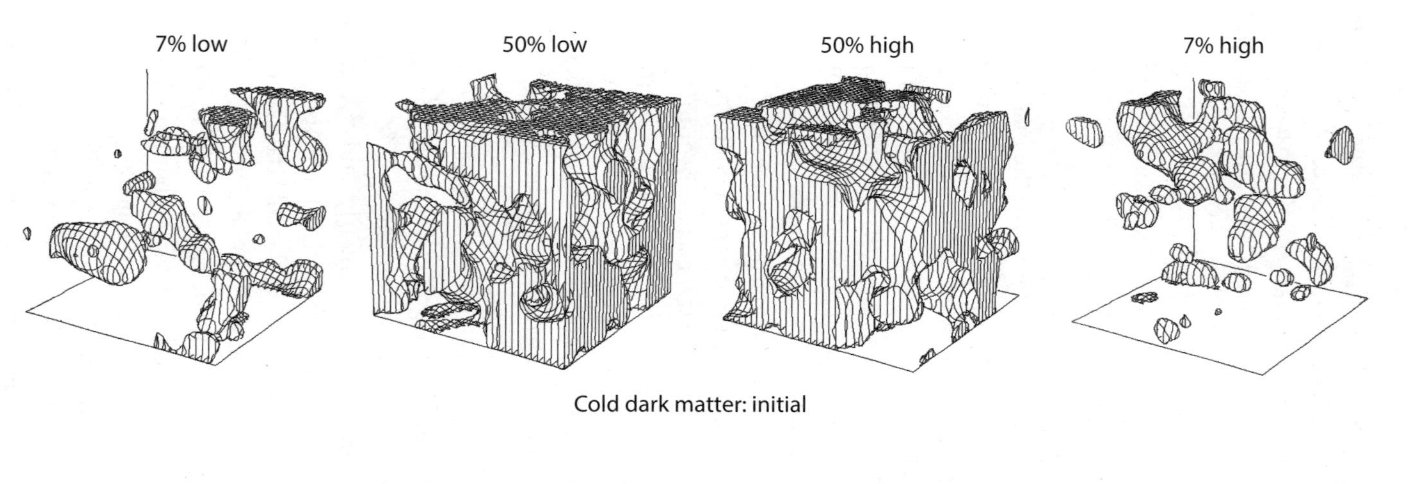

Cold dark matter: initial

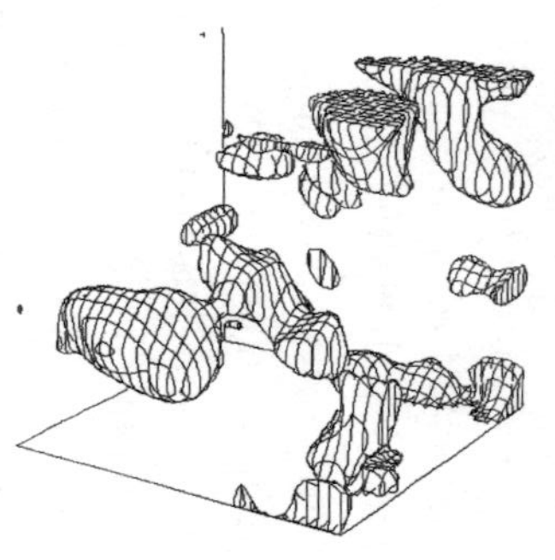

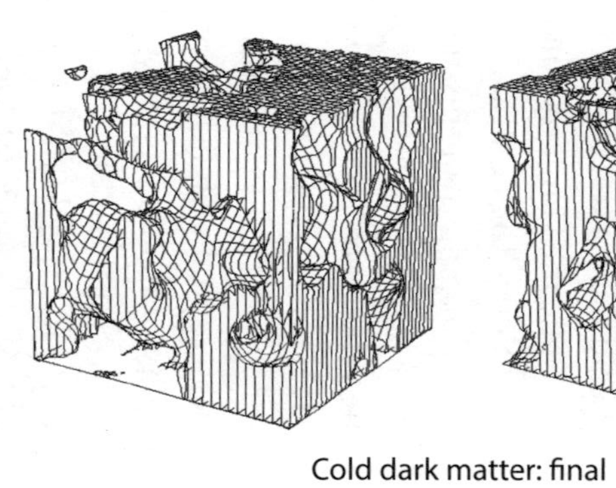

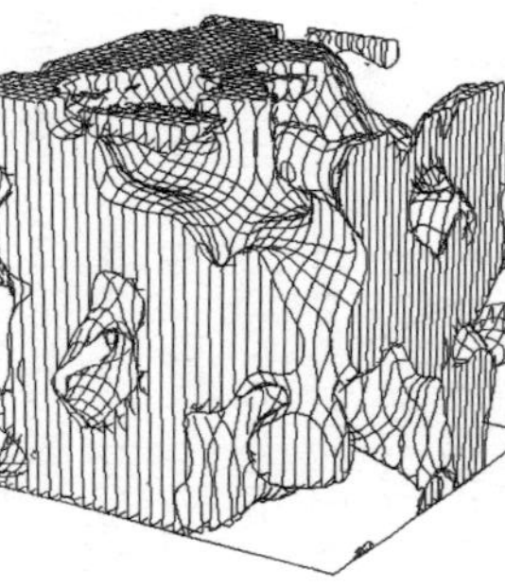

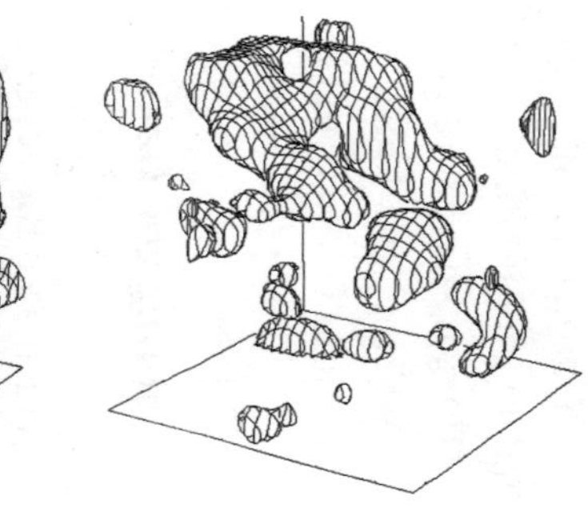

Cold dark matter: final

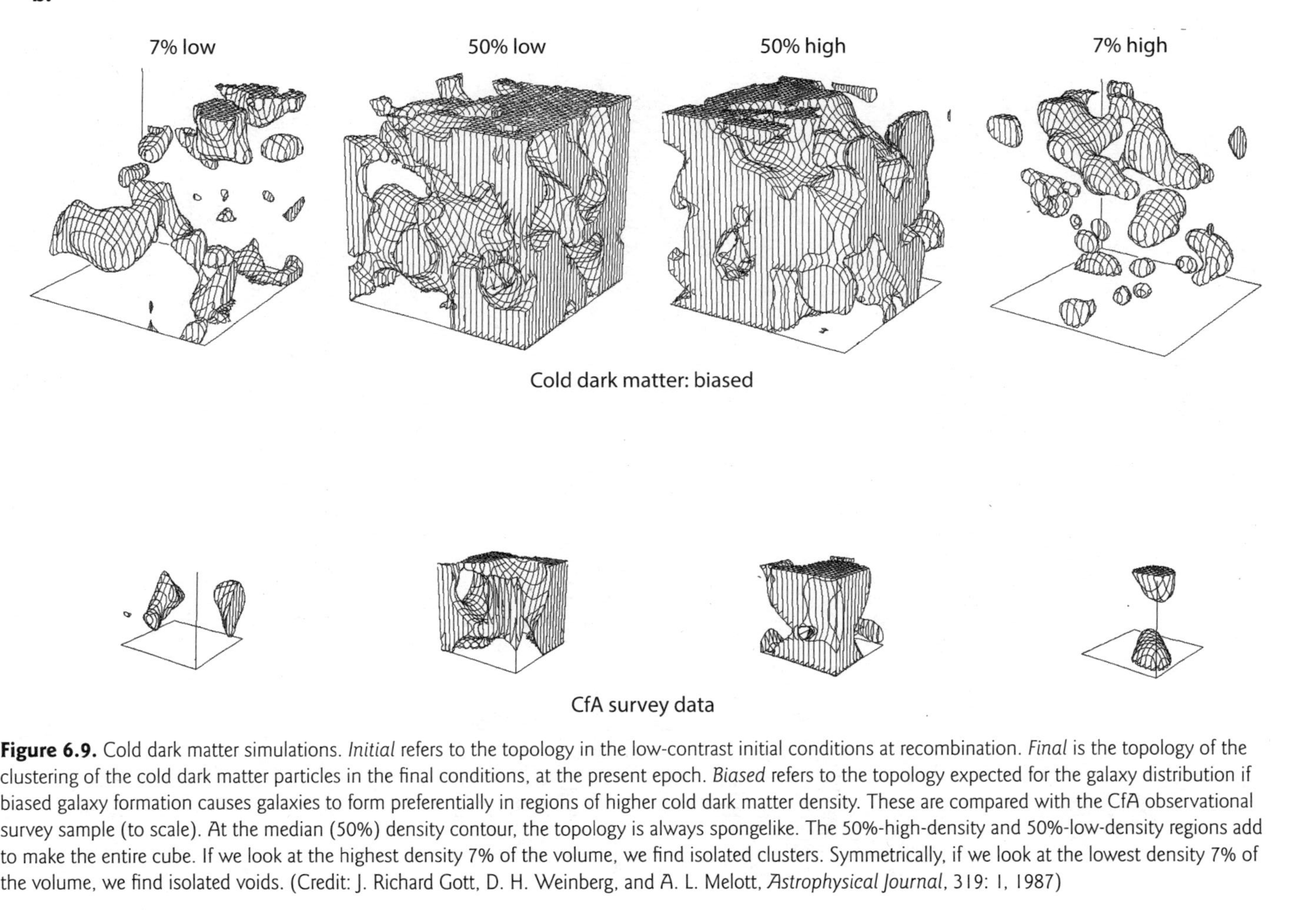

Figure 6.9. Cold dark matter simulations. *Initial* refers to the topology in the low-contrast initial conditions at recombination. *Final* is the topology of the clustering of the cold dark matter particles in the final conditions, at the present epoch. *Biased* refers to the topology expected for the galaxy distribution if biased galaxy formation causes galaxies to form preferentially in regions of higher cold dark matter density. These are compared with the CfA observational survey sample (to scale). At the median (50%) density contour, the topology is always spongelike. The 50%-high-density and 50%-low-density regions add to make the entire cube. If we look at the highest density 7% of the volume, we find isolated clusters. Symmetrically, if we look at the lowest density 7% of the volume, we find isolated voids. (Credit: J. Richard Gott, D. H. Weinberg, and A. L. Melott, *Astrophysical Journal*, 319: 1, 1987)

Thus we can measure the topology in the smoothed galaxy distribution today, and deduce the topology of the initial conditions, and compare it to the predictions of inflation. At the bottom, we show our small cubical CfA observational sample at the same scale and with the same smoothing length. It displays a spongelike median density contour (the 50% low—50% high pictures) and isolated voids at the 7% low contour and isolated clusters at the 7% high contour. It also shows a similar number of structures per unit volume as the Cold Dark Matter simulations.

Figure 6.10 (top graph) shows the genus curve for the cold dark matter initial conditions, the cold dark matter final conditions, and the final biased conditions—tracing where we expect to find the galaxies. As expected from Figure 6.9, at the 7% low-density contour, we find isolated voids, at the median (50%) contour surface, we find a spongelike topology, and at the 7% high-density contour, we find isolated clusters. The genus in Figure 6.10 (top) starts out negative (isolated voids), becomes positive (spongelike genus) in the middle, and becomes negative again (isolated clusters) at the end. For comparison, the theoretical curve is shown. The fit is very good, within the errors.

Then we measured again our small CfA observational sample, which included the Virgo Supercluster. Its median density contour had four holes. It was spongelike. It had two isolated high-density clusters and two isolated low-density voids. With twice as many holes as clusters or voids, its observed genus curve couldn't have possibly looked more like the theoretical curve, given the small number of structures we were measuring (see Figure 6.10, bottom graph).

Simulations of Swiss Cheese Universes

We could also calculate genus curves for the Zeldovich model. The genus curves for a Voronoi honeycomb look quite different. Here we placed 8 points at random in our simulation box and constructed a Voronoi honeycomb with these as the 8 cell centers. We placed all the galaxies on the cell walls. This was the Zeldovich model. Then we smoothed the data and constructed density contour surfaces (Figure 6.11).

One can see at a glance that this produces a Swiss cheese topology. Just look at the 50% high picture, showing the parts that are higher than

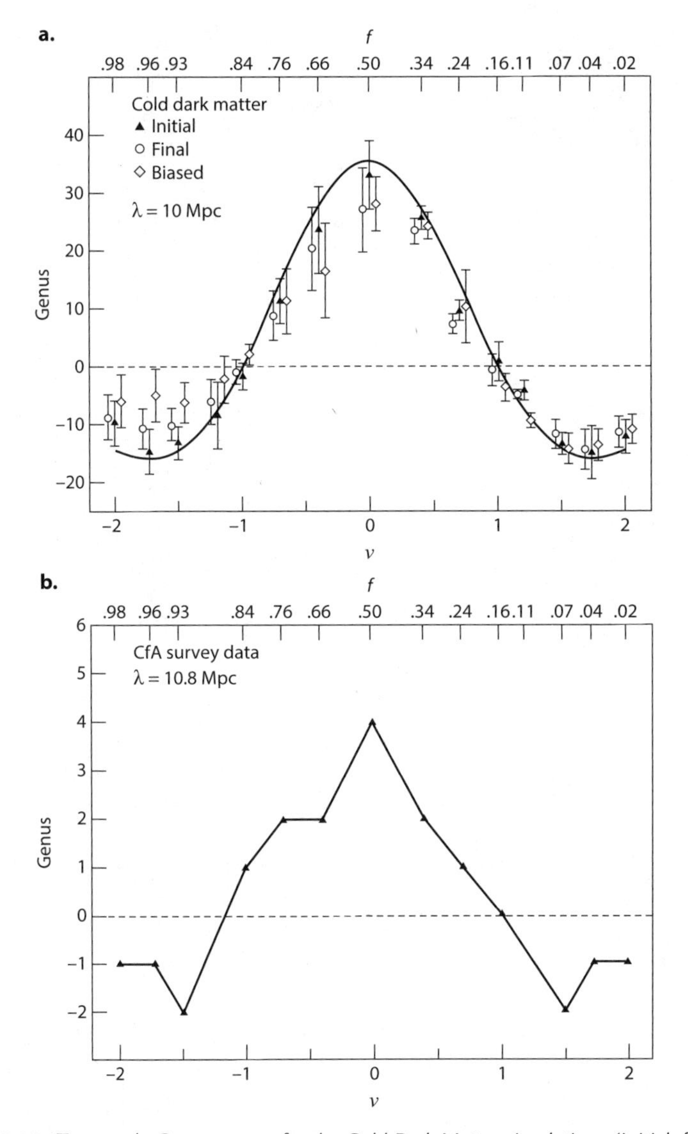

Figure 6.10. *Top graph*: Genus curves for the Cold Dark Matter simulations (initial, final, and biased; see Figure 6.9) are shown in the top graph. Mean values with error bars are shown. The smooth solid curve is the genus curve predicted for Gaussian random-phase initial conditions expected from inflation; *f* is the fraction of the volume on the high-density side of the contour surface being measured. *Bottom graph*: Genus curve for the observational CfA sample. It is only a small sample but still has approximately the same shape as the genus curve for Gaussian random-phase initial conditions shown in the simulations above. (Credit: J. Richard Gott, D. H. Weinberg, and A. L. Melott, *Astrophysical Journal*, 319: 1, 1987)

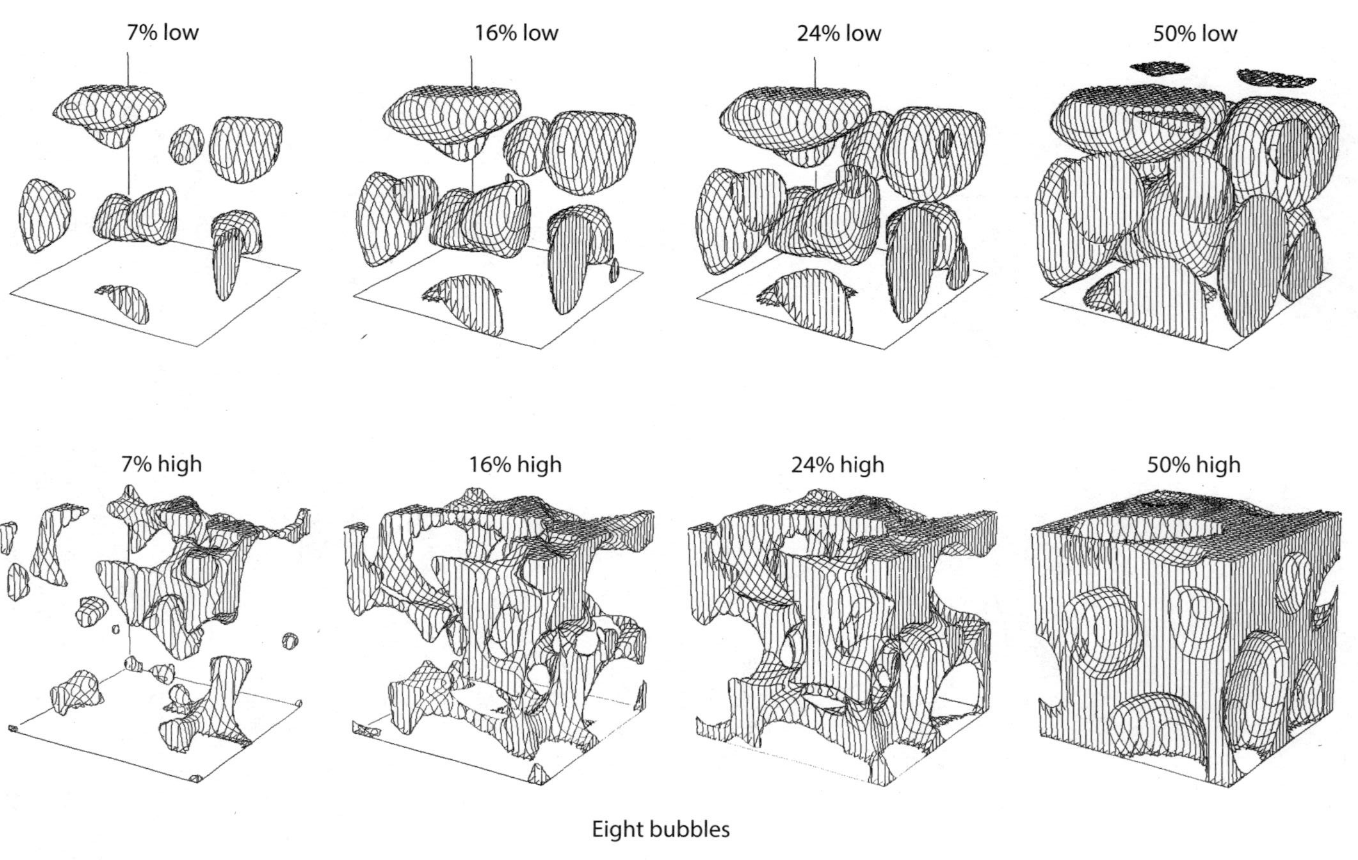

Figure 6.11. A Voronoi honeycomb with 8 cells (or bubbles) was created to simulate a Swiss cheese universe. The topology of the simulation, with appropriate smoothing, at different density thresholds is shown. The 50%-high-density picture looks very much like a block of Swiss cheese. (Credit: J. Richard Gott, D. H. Weinberg, and A. L. Melott, *Astrophysical Journal*, 319: 1, 1987)

the median density—it looks exactly like a block of Swiss cheese.[5] The 50% low picture shows the low-density voids—the low-density Voronoi cells. As we go to still-lower density contours, the 8 low-density voids are still seen intact in the 24% low, 16% low, and 7% low pictures. If we continue to higher-and-higher-density contours (above the median), we see high-density filaments where the cell walls meet, in the 24% and 16% high pictures, just as in the Zeldovich model, and, finally, at the highest density, in the 7% high picture, we see isolated clusters where the filaments meet. The genus curve corresponding to these pictures (Figure 6.12) is quite different from the symmetric curve seen for a sponge-like topology.

The left hand of the genus curve is flat at the value of genus = −8, as it identifies the 8 low-density voids created by the Voronoi Honeycomb.

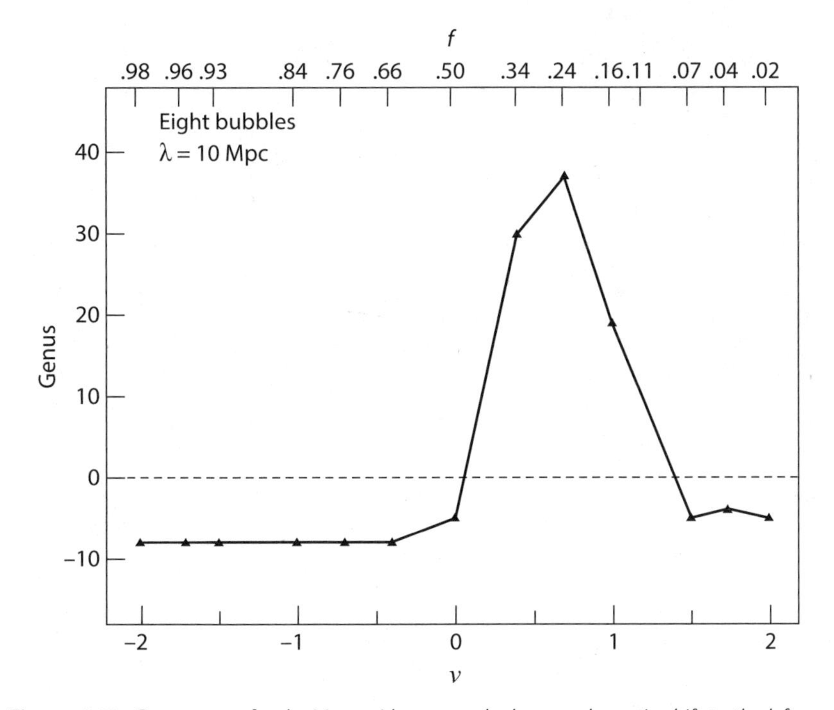

Figure 6.12. Genus curve for the Voronoi honeycomb shows a dramatic shift to the left—a Swiss cheese shift. The genus value is −8 and constant for all low-density contours (on the right) showing 8 isolated voids. The median density contour $\nu = 0$ is negative, due to isolated voids. This does *not* look like the genus curve for the observations (Figure 6.10, bottom graph). (Credit: J. Richard Gott, D. H. Weinberg, and A. L. Melott, *Astrophysical Journal*, 319: 1, 1987)

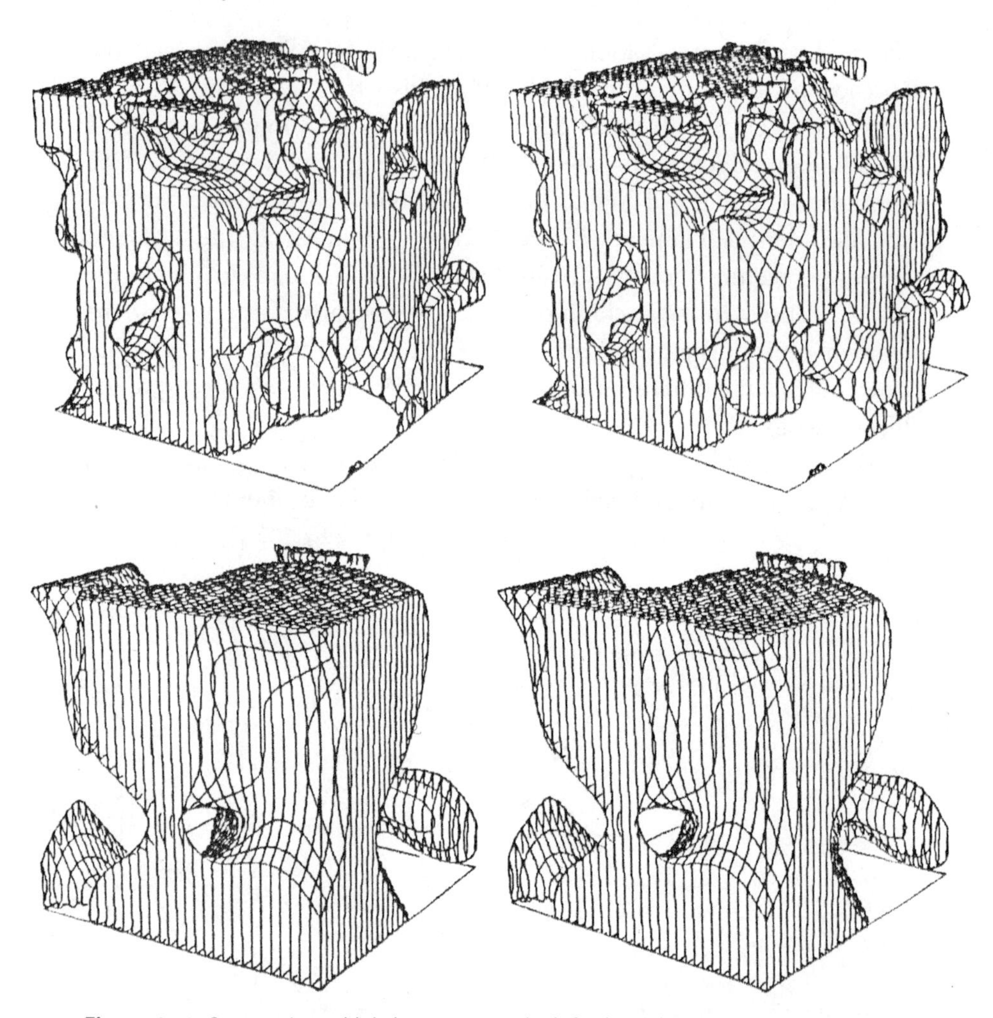

Figure 6.13. Stereo pairs: cold dark matter–50%-high final simulation (*top pair*), and the 50%-high CfA observational sample (*bottom pair*). To observe in 3D, touch your nose to the page. The left-eye image of cold dark matter–50%-high pair will be directly in front of your left eye, and the right-eye image will be directly in front of your right eye. They will fuse into one blurry image. Slowly move your head back, keeping the image fused. As you move your head back, the fused image in the center will come into focus—in 3D! It takes some practice to do this—give it a try. Move down and repeat the procedure to see the 3D image for the CfA data. (Credit: Adapted from J. Richard Gott, David H. Weinberg, and A. L. Melott, *Astrophysical Journal*, 319: 1, 1987)

Figure 6.14. David Weinberg (*far left*) and the author view slides of the topology in 3D using red-blue glasses and discuss the topology results (*right*) in 1987. (Credit: Photos courtesy of Princeton University, Bob Matthews)

The genus at the median contour (in the center of the graph) is negative, showing a Swiss cheese topology at the median density contour. At higher-density contours, the genus becomes positive and the topology becomes spongelike (Zeldovich's network of filaments at the edges of the cells) before it finally breaks apart into isolated clusters at very high density (at the corners of the cells). This shifts the genus curve radically to the left—what we call a Swiss cheese shift. The genus curve for the observations (Figure 6.10, bottom) does not look like the Swiss cheese model at all: the observations show a symmetrical genus curve, which is spongelike at the median density contour in the center.

Finally, we show two stereo pairs (Figure 6.13): the cold dark matter (CDM) simulation, final conditions; and the original cubical CfA observational dataset (adapted from Gott, Weinberg, and Melott 1987). In both cases we show the high-density halves. Both are sponges, where the high-density half is in one multiply connected piece, full of holes, allowing the low-density half to percolate through it.

We could view these 3D images by projecting a red left-eye image and a blue right-eye image on the screen and viewing them with red-blue 3D glasses, as used in 1950s' sci-fi movies (see Figure 6.14).

Einasto Weighs In

As he relates in his 2014 book *Dark Matter and the Cosmic Web Story*, when Einasto saw our Gott, Melott, and Dickinson (1985) preprint on the spongelike topology of the universe during a visit to the European Space Agency, he realized that he had made an error in not writing a paper attempting to characterize the topology of the distribution. He set out immediately to do so. He used a code to measure the maximum size of connected systems. He set density thresholds for the voxels in a computer simulation. He found that at medium-density thresholds, the low-density voxels formed a connected system whose size was equal to the simulation box size, as did the high-density voxels. That supported the spongelike topology we were claiming. In the observations, since he was not smoothing the data, he could not study the connectedness of voids—empty voxels always filled the volume, but he did find that the high-density voxels (containing galaxies) produced connected structures that extended completely across the observational box as one lowered the density threshold to medium densities. He rushed out a preprint on this. It supported our spongelike topology and reflected the view he expressed at the 1983 Crete Symposium (which I attended), where he emphasized filaments connecting clusters. I was quite happy to see this preprint; it provided some unexpected support from the Soviet school. We were able to add this preprint to our references in the final version of our paper, which appeared in the *Astrophysical Journal* in 1986.

On November 9, 1986, the *New York Times* covered our paper in a page 1 story titled "Rethinking Clumps and Voids in the Universe," written by the eminent science writer James Gleick. This article included a kind quote from Jerry Ostriker: "It's a clever new approach. It looks to be a powerful tool for discriminating between different physical models for how the universe got its structure, and that's the really exciting thing."

We were now eager to test our theory on larger data sets.

Chapter 7

A Slice of the Universe—the Great Wall of Geller and Huchra

When I was at Caltech, I met Ed Turner, a graduate student working on his thesis on binary galaxies. I found Ed's work with binaries especially interesting because of my own interest in groups of galaxies. We decided to team up and make a catalog of groups of galaxies, using the Zwicky catalog of galaxies down to magnitude 14. This included galaxies up to 1,600 times fainter than the faintest naked-eye stars. We wrote a computer program to find groups of galaxies by locating regions of the sky where the counts of galaxies per square degree were statistically above average. This was one of the first instances of using a computer program to pick candidates rather than doing it by hand. One knew the rules the computer had used and, therefore, what the results meant. We found about a hundred groups. Many of these were well known, like the famous Virgo cluster, but many were new.

Many of these groups already had redshifts for a few of their galaxies taken by observers. As we have seen, the redshift of a galaxy is the shift in the galaxy's spectral lines toward the red, indicating its recessional velocity from the earth. This is proportional to the galaxy's distance, according to Hubble's law. But, in addition, galaxies can have *peculiar*, or individual, velocities due to their orbital velocities within clusters. For galaxies within an identified group, there would be a mean velocity of recession for the group as a whole, plus random velocities associated with the orbital velocities of the galaxies within the group. Having the orbital velocities of the galaxies within a group and knowing the size of the group, we could determine the group's mass. Overall, we found the groups' mass-to-light ratios to be about 140 solar masses per solar luminosity, similar to but a bit smaller than what Zwicky had found for the Coma cluster (about 500 solar masses per solar luminosity). This was understandable, since the Coma cluster contained many galaxies whose new-star formation had stopped, making them less bright for their mass than most spiral galaxies in small groups where star formation was still going on. We found a lot of dark matter, as Zwicky had, but definitely not enough to cause the universe to recollapse in the future: that would require 1,500 solar masses per solar luminosity. We concluded that the mass associated with groups of galaxies was not enough ever to halt the overall expansion of the universe.

Our group catalog provided an opportunity for young observers. Alan Sandage of Palomar Observatory had taken the redshifts of a multitude of galaxies for his many projects; it was hard to compete with him. But a young observer could pick some of our groups that had no measured galactic redshifts, measure them all, and determine mass-to-light ratios for these groups. Bob Kirshner did just this, measuring the redshifts for galaxies in six of our groups. He found a mass-to-light ratio for these groups of 204 ± 60 in solar units. Then Marc Davis, David Latham, and John Tonry built a sensitive spectrograph (following a design pioneered by Steve Shectman), which could be used on the 1.5-meter-diameter telescope at Mount Hopkins to turn it into a redshift-measuring machine for galaxies. By 1983, Davis, John Huchra, and Tonry had used the Zwicky catalog as a finding chart to measure the redshifts of about 2,400 galaxies. Their survey included galaxies as faint as magnitude 14.5, about 1.6 times fainter than the faintest galaxies in our group catalog. Davis and Jim Peebles used the new redshift catalog to statistically estimate mass-to-light ratios. They got a mass-to-light ratio of about 300 in solar units, about ⅕ of that required to eventually halt the expansion of the universe, corresponding to $\Omega_m = 0.2$. When Davis left Harvard to go to Berkeley, Huchra and Margaret Geller inherited the redshift-measuring machine. I knew them both well. John Huchra had been a graduate student at Caltech when I was there as a postdoc. He took a long time to finish his thesis because he had notoriously bad luck with cloudy weather on his observing runs.

He would make up for this later in his career by spending more clear nights observing than almost any astronomer in his generation. Margaret Geller had been a star graduate student in physics at Princeton when I was there as a graduate student in astrophysics. She had been recruited from Berkeley by John Wheeler himself, the father of black hole physics and Richard Feynman's mentor. Wheeler suggested she work on flattened neutron stars for a thesis, but Margaret decided to work with Jim Peebles on galaxy clustering instead. At Harvard, Margaret and John started a new, deeper survey to magnitude 15.5. They included galaxies down to about 2.5 times fainter than the ones Davis and his colleagues had surveyed. Since brightness falls off like the square of the distance, this meant that they were surveying out to a distance about $\sqrt{2.5}$, or 1.58, times further away. But they couldn't study as large an area of the sky right away—that would be too many galaxies—so they chose to do a thin slice first.

A Thin Slice

In talks, Margaret famously demonstrated the reasoning for doing a slice. She would get out a big map of the world and cut out a little square covering about 19° on a side. Picking a square region at random, you might land in an ocean, on a continent, or on a little of both. If that were all you surveyed, you might mistakenly think Earth was all ocean or all land. But then she took her scissors and cut out a thin band, 1° wide, stretching 360° around Earth's equator. This thin band had the same area as the small square, yet it really showed you what the topography of Earth was like. As you moved along the equator, you would encounter Africa, the Atlantic Ocean, South America, the Pacific Ocean, Indonesia, and the Indian Ocean. Your sample would inform you about the nature and extent of continents, and you could estimate the fraction of land and water. This same reasoning applied when considering how to sample the sky, so Margaret and John, together with graduate student Valerie de Lapparent, made a deep slice of the universe, covering a thin band in the sky about 6° wide and about 120° long.

In three dimensions, this thin-slice survey looked like a Japanese fan with Earth at the apex. You could take this fan and lay it flat on the page of

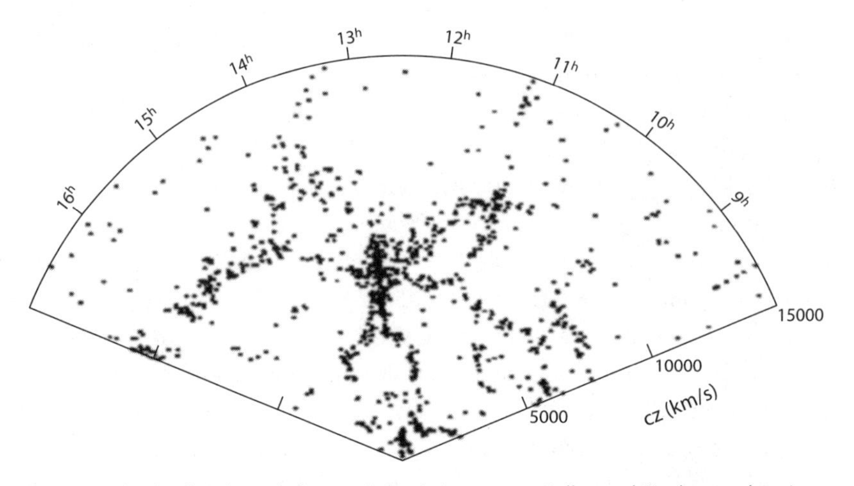

Figure 7.1. In this fan-shaped slice made by de Lapparent, Geller, and Huchra, each point represents a galaxy. Earth is at the bottom vertex of the fan. The upper curved edge of the fan shows galaxies receding from us at 15,000 kilometers per second at a distance of 730 million light-years. Notice the thin filaments and large empty voids. (Credit: V. de Lapparent, M. J. Geller, and J. P. Huchra, *Astrophysical Journal*, 302: L1, 1986)

a book to make a fan-shaped map of the universe. Their map, published in the *Astrophysical Journal* in 1986, was quite startling (Figure 7.1).

Earth was located at the vertex of the map at the bottom. Radial distance from Earth is measured using the recessional velocities of the galaxies. With Hubble's law ($d = v/H_0$), we just have to divide these recessional velocities by Hubble's constant to get the distances to the individual galaxies. The most distant galaxies in the survey were receding from us at a velocity of 15,000 kilometers/second. With a value of H_0 = 67 kilometers/second/megaparsec, we find that the most distant galaxies in the survey are at a distance of 224 megaparsecs, or 730 million light-years. Note the scale on the right, with markers for recessional velocities of 5,000, 10,000, and 15,000 kilometers/second. Around the outside of the fan the celestial longitude is marked off in hours of *right ascension*. Astronomers divide the 360° of longitude on the celestial sphere into 24 hours of right ascension, each taking up 15°. In the center of the map, vertically straight above the bottom of the fan, at a recessional velocity of about 7,000 kilometers/second, is the Coma cluster. It looks like a vertical dagger of dots, each dot representing a galaxy. The Coma cluster is compact. As a whole it is receding from us at 7,000 kilometers/second, but its galaxies have individual motions of ±1,000 kilometers/second along our line of sight due to their orbital motions within the cluster. These orbital motions spread out their recessional velocities over a range from about 6,000 to 8,000 kilometers/second. This makes the Coma cluster a "finger of God" pointing directly at Earth. If the cluster has a crossing time 10 times shorter than the age of the universe, this "finger" on the redshift map is elongated by a factor of 10 to 1 in the direction of Earth. This shows the cluster is gravitationally bound—which Zwicky understood. Jim Gunn and I had noticed this effect when we were working on the Coma cluster, so this view of the Coma cluster was not a surprise. But the rest of the galaxies appeared to be laid out on narrow strings wandering in different directions. Very large, nearly empty voids, surrounded by thin walls of galaxies, were visible. On the left was an elliptical void next to the Coma cluster, having a width of about 5,000 kilometers/second—about the same width as the great void in Boötes. Because this elliptical void seemed to have curved edges, Margaret and John likened it to a bubble, and they likened the whole distribution to a froth of bubbles. (Bubbles allowed one to conceive of voids as blast-wave explosions that snowplowed up material at their borders.) Other voids of various shapes appeared. I first saw this picture at a Tuesday Lunch at the Institute of Advanced Study, where a copy was passed around.

Most people looked at this picture and pronounced the Swiss cheese universe of Zeldovich triumphant. A froth of bubbles was equivalent to a Voronoi tessellation, which was equivalent to a honeycomb with cells, which was equivalent to Swiss cheese in topology. In all these cases, isolated voids were completely surrounded by high-density galaxy regions. Such a froth of bubbles, in which the galaxies lived on the bubble surfaces, had isolated low-density bubble interiors. You could travel along one bubble surface to the next without ever entering the empty voids; in other words, the high-density part was all in one piece. If you looked at a slice through such a froth or honeycomb, you might expect to see empty cells surrounded by a grid of walls.

But opinion was not unanimous. Our spongelike results were supported in a 1987 *Astrophysical Journal* paper by White, Frenk, Davis, and Efstathiou, who compared their *N*-body simulations with Geller,

Huchra, and de Lapparent's new CfA (the Harvard-Smithsonian's Center for Astrophysics) sample. They said: "The observations show a rather wide range of morphologies. Some regions appear to show a 'sudsy' structure like that emphasized by de Lapparent, Geller, and Huchra, whereas others appear less organized. . . . The galaxy distribution in our models appears to us to have a spongy topology. Voids do not seem surrounded by continuous unbroken sheets of galaxies, but rather percolate through the computational volume." They recommended a quantitative study of the topology of the observations and simulations and quoted our Gott, Melott, and Dickinson (1986) paper.

A Thicker Slice

Geller and Huchra continued to take data and made two more slices, each 6° wide, on the top and bottom of the first one. These combined to widen the original slice to 18°. In the first slice, the features that attracted attention were the voids with the thin strings of galaxies separating them, but in this new thicker slice, while the original voids were still visible, something new began to dominate the scene.

I first caught sight of this "Great Wall" of Geller and Huchra at a meeting in Rio de Janeiro in 1989. Experts on the large-scale structure of the universe had gathered there from all over the world. It was a dramatic setting with Sugarloaf Mountain in the bay and the famed statue "Christ the Redeemer" atop Corcovado Mountain overlooking the city. Our hotel was right on Copacabana Beach. I was there to talk about the spongelike topology of large-scale structure. The assembled astronomers had all seen the slice of the universe from de Lapparent, Geller, and Huchra. That was a 120°-wide fan that was 6° in thickness. It showed the voids that had become famous. People expected this to be the centerpiece of the conference discussion, but we were in for a surprise. Huchra could not come to the meeting, but he had given Ed Bertschinger the latest results to show. If the de Lapparent, Geller, and Huchra 6°-thick slice was a nice slice of pizza to eat, this new 18°-thick slice was a slice of deep-dish pizza—thick enough to savor.

Bertschinger had brought a slide of the new, thicker slice. He did not have copies to hand out, and this was the first time any of us had seen it.

I took a picture of the slide on the screen—it was that memorable. Visually, the original 6°-thick slice had the appearance of individual voids outlined by a lacework of walls. The thicker slice revealed that some of the cell walls in the original picture had become more prominent than others to make one enormous, continuous structure 758 million light-years long spanning the entire width of the picture: a giant wall, dubbed the Great Wall, with echoes of the Great Wall of China stretching across central Asia.

For the audience at the conference, the original slice was surprising enough, but this Great Wall was an even more dramatic revelation.

This large structure challenged our idea that the universe was homogeneous at very large scales. Anyone looking at the original thin-pizza slice of the universe would pronounce the giant voids the most prominent feature. But most remarkable feature in this deep-dish pizza slice was that one Great Wall filament snaking across the entire survey. The universe was supposed to be clumpy on small scales, becoming smoother and smoother as one looked at larger and larger scales. On the largest scales, the cosmic microwave background was uniform to at least one part in 10,000, reflecting this idea. So what was going on? Also, why was it a linear feature anyway? If individual bubbles were forming and hitting each other to produce a froth, then why did the walls of a number of separate independent bubbles line up and strengthen along a single line to form a Great Wall? Could that happen by chance? (See Color Plate 4.)

Speculations abounded for the rest of the conference as we all tried to make sense of what we had seen. Cosmic voids surrounded by walls might be made by nuclear explosions, as Ostriker had proposed, but this was different—this was one long wall passing several adjacent voids. What could make that? I remember Marc Davis saying the Great Wall didn't look like it was made by normal gravitational accretion in the standard cold dark matter cosmological model. Nothing that dramatic had shown up in his group's computer simulations.

I also recall Davis tossing out the idea that the Great Wall could be a cosmic string wake. Cosmic strings are thin strands of high-density vacuum energy, not yet observed but predicted by theories that unified forces in the early universe. The theory of inflation predicted that in the early universe, a very high-density vacuum state permeated all of space.

It was like a uniform, snow-covered field. But eventually the energy of the high-density vacuum state decayed into thermal radiation as inflation ended. This caused a melting of the high-density quantum vacuum state. Cosmic strings are like the snowmen left standing after the snow has melted. As the high-density vacuum state decays, some high-density vacuum from the early universe remains trapped in cosmic strings. These high-density threads of vacuum energy would be narrower than an atomic nucleus. Cosmic strings have no ends, so they must occur either as finite closed loops or as infinite strings stretching across the visible universe. Think SpaghettiOs plus long strands of spaghetti. The strings Davis was talking about using to explain the Great Wall were the infinite ones. These strings are under tension, like rubber bands, and would be whipping around at high velocities, approaching a fraction of the speed of light.

No one in the audience should have been more excited than I to hear talk of cosmic strings—since I had worked on them. In 1985, I and William Hiscock independently discovered an exact solution to Einstein's field equations for the geometry around a cosmic string. The geometry of this solution implied that if a rapidly moving cosmic string were to pass directly between two gas clouds, the gas clouds would be drawn together and collide. This is the idea Davis was talking about. As the cosmic string plows at high velocity through intergalactic gas, it will cause gas to collide in a sheet in its wake. Gas collision causes dissipation, increases density, and can trigger galaxy formation. If a string is plowing along like the keel of a boat, a wake of newly formed galaxies can be left behind it. Could the Great Wall of Geller and Huchra be the galaxies formed in the wake of a moving cosmic string? It seemed as if something as dramatic as a cosmic string might be required to produce a structure as spectacular as the Great Wall.

In his conference summary (see Davis 1991), Marc Davis noted that "there are substantial gaps in the plane of the structure so that the overall topology appears spongy," supporting our results, but he added: "By visual inspection, the CfA [Geller and Huchra] filaments appear larger and more prominent than in any available CDM [cold dark matter] *N*-body simulation."

For my part, I thought that the gravity of cold dark matter alone might be able to make this structure. I expected a spongelike geometry of galaxy clustering, with high-density filaments connecting galaxy clusters. If you could go from cluster to cluster by traveling along high-density filaments in the sponge, you could imagine working your way all across the universe. Maybe one prominent long filament connected across the entire sample was something the standard inflationary Big Bang initial conditions could produce. To find out, one would really have to do a large computer simulation, and no one had yet done a computer simulation large enough to encompass the entire volume containing the Great Wall.

Chapter 8

Park's Simulation of the Universe

I couldn't wait to get home to Princeton after hearing in Rio about the Great Wall of galaxies. I told my graduate student at that time, Changbom Park (currently Professor at the Korean Institute for Advanced Study), that he should stop what he was doing and devote full time to making the world's largest computer simulation. Time was short because everyone had seen that picture. Our department chairman, Jerry Ostriker, had just bought our department a supercomputer—so doing a new, largest simulation was possible. In only three weeks, Changbom wrote an *N*-body code to do the calculation. This was remarkably fast. The new simulation would involve 4 million particles. In 1975 Sverre Aarseth, Ed Turner, and I had created a 4,000-body simulation of galaxy clustering in the universe, breaking the existing record. Changbom would publish his thesis in 1990; thus, in 15 years, the number of particles had increased by a factor of 1,000. We joked that in another 15 years (by 2005), we should be doing 4 billion particles! (It seemed fantastic at the time, and yet Volker Springel and his colleagues would actually do a 10-billion-particle simulation by 2005.) Our extrapolation was consistent with Moore's law, the empirical observation that computer power seems to double every 1.5 years. After 4.5 years (three doublings) we should see an increase of a factor of 8. Five years gives an increase of about a factor of 10, and in 15 years, if it continues in the same way, that should be three factors of 10, or a factor of 1,000. Perhaps no field has benefited more from the computer revolution than astronomy. Computers allow us to analyze ever-larger observational samples, and they also allow us to do ever-larger simulations to model them.

Changbom would be using Jim Peebles's theory of cold dark matter and the Big Bang inflationary model as the basis for his initial conditions. He would lay down small-density perturbations as waves. Recall our analogy of ocean waves, where a wave crest represents a region of slightly higher than average density and a wave trough represents a region of slightly lower than average density. A single wave would be sinusoidal, with crests and troughs lined up as parallel furrows. The wave would have an amplitude (in density), a wavelength (distance between successive wave crests), and a direction. Changbom was going to lay down the waves at random. Each wave would be given a random direction and a random phase; that is, the wave crests and troughs would point in a random direction, and they would be shifted a random distance along the direction of the wave, so that a particular point would be equally likely to be part of a wave crest or a wave trough. The cold dark matter theory coupled with inflation would tell Changbom the mean-square amplitude of the waves at a given wavelength; then he would pick the amplitude of the particular wave randomly from a bell-shaped (Gaussian) distribution with that mean-square amplitude. This gives what is called a *Gaussian random-phase distribution*—the initial conditions predicted by inflation. This distribution gives a spongelike topology of high- and low-density regions in the initial conditions, as we have discussed. Inflationary theory tells us how much power (square of the amplitude of the waves) there is on average in fluctuations as a function of their wavelength. We call this the *power spectrum*. Just as a spectrum gives us information on how much light is present as a function of wavelength, the power spectrum tells us how much power there is on average in density fluctuations of different wavelengths. Inflation predicts the amplitude of fluctuations on different scales.

Using Peebles' cold dark matter theory, one could follow the growth of the fluctuations through the action of gravity up through the period of recombination and after. Fluctuations then grew by gravity. Using this theory as a basis, Changbom was able to start his simulations when the universe was about a factor of 24 smaller than it is today. All the density fluctuations at this point were small with respect to unity (i.e., a few percent), and they were already growing due to gravity. Two million matter particles were placed on a 128 × 128 × 128 cubical grid and displaced slightly to produce the density fluctuations. They were given the individual initial velocities that they would have attained due to the growth of those fluctuations. Changbom assumed a density of matter that was 40% of that required to ever halt the expansion of the universe, close to the value of 30.8% measured by the Planck Satellite Collaboration in 2014. Most of this matter is in the form of cold dark matter, with the rest contributed by ordinary atoms. The computer then calculates the movement of these matter particles due to their mutual gravitational interactions. Essentially, Isaac Newton takes over. Slightly overdense regions become denser still as their excess gravity draws other particles in, while underdense regions grow into voids. The computer follows the complicated nonlinear process as clusters of galaxies and dense filaments form. The computer program used a fast Fourier technique to cut the computation time in calculating the gravitational forces on the cold dark matter particles. This technique had been pioneered by R. W. Hockney and J. W. Eastwood, Kevin Prendergast and R. H. Tomer, George Efstathiou, and Adrian Melott, among others. Standard computations of gravitational forces (such as those Peebles and Aarseth used) required of order N^2 operations per time step, whereas the fast Fourier technique required only of order $N \ln N$ operations. For $N = 2$ million, for example, $N^2 = 4 \times 10^{12}$, whereas $N \ln N = 2.9 \times 10^7$, for a considerable savings.

In the initial conditions, Changbom also introduced a second, biased set of 2 million particles placed to represent locations where individual galaxies were likely to form. It is recognized that galaxies are more likely to form in regions of higher cold dark matter density. The ordinary matter tends naturally to fall into such regions and form galaxies there. Likewise, galaxies are even less likely to form in voids where the matter density is low. (Observationally, today we find fluctuations of order unity in the number of galaxy counts in spheres of radii 39 million light-years, whereas fluctuations in the dark matter density on the same scale are only 80% as large.) So, in regions of higher-than-average matter density in the initial conditions, more "galaxy" particles were placed, and in regions of lower-than-average matter density fewer galaxy particles were placed. James Bardeen and his colleagues had developed formulas for doing this. These galaxies were then allowed to move due to the gravitational influence of the matter particles. As the matter particles formed clusters, the galaxy particles fell into them. As the universe simulation expanded by a factor of 24 to reach the present epoch, the full range of galaxy clustering occurred.

The simulation was divided into 256 × 256 × 256 cells on which the gravitational field was calculated. The boundary conditions on the box were periodic, front matching the back, left side matching the right side, and top matching the bottom, so Changbom was effectively modeling an infinite universe made of identical stacked boxes. No place in the cubical box was special. This is fine for modeling structures smaller than the cubical box size. The whole simulation modeled a cube 1.49 billion light-years on a side at the present epoch, large enough to encompass structures as large as the Great Wall. But would such a structure appear?

Finding a Great Wall in the Simulation

We were quite excited to look at the first pictures. Changbom just sliced the cube in thick slices like a loaf of bread—and there it was, a structure just as magnificent as the Great Wall! We found it right away looking at the 3D sample. The next step was to produce a simulated fan-shaped survey that would mimic the original 6°-thick fan published by de Lapparent, Geller, and Huchra. We simply placed an "Earth" in our simulation box and considered how the universe would "look" from that vantage point. We surveyed galaxies in a fan-shaped slice of the universe of the same size and shape as they had observed. Changbom gave the simulated galaxies brightnesses taken from the distribution found in the real universe and included only the simulated galaxies that would have been bright enough for de Lapparent, Geller, and Huchra to have seen from this Earth. The simulations are compared with the real de Lapparent, Geller, and Huchra slice in Figure 8.1.

The agreement was extraordinary! Both pictures showed large, empty voids surrounded by thin filaments and a prominent filament, or wall, extending from left to right all the way across the survey. This agreement was not the result of any extensive trial-and-error search through the data cube. We simply found a long filament immediately in the 3D data cube, placed an observation point at approximately the right distance from it, and took a snapshot picture.

The simulation also includes a great cluster, just like the Coma cluster, which appears as a sharp dagger pointed at Earth, sitting at the right-hand end of the simulated Great Wall. In the de Lapparent, Geller,

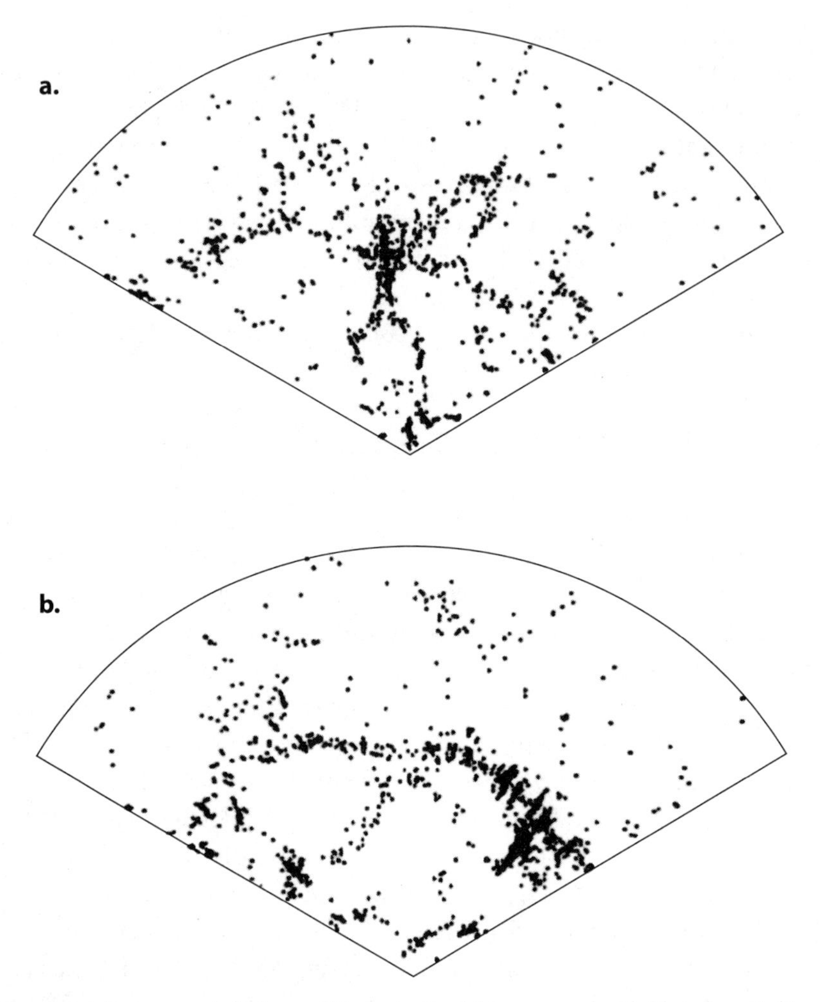

Figure 8.1. De Lapparent, Geller, and Huchra's slice (a) is compared with Changbom Park's simulated slice (b). Park's computer simulation using Cold Dark Matter and biased galaxy formation produced a Great Wall and large voids remarkably like those seen in the de Lapparent, Geller, and Huchra observational slice. (Credit: Changbom Park, *Monthly Notices of the Royal Astronomical Society*, 242: P59, 1990)

and Huchra slice, the Coma cluster of galaxies appears as a comparable downward-pointing dagger in the Great Wall right at the center of the fan. We also made thicker slices, and these showed Changbom's simulated Great Wall even more dramatically—just as occurred in the case of the Great Wall of Geller and Huchra.

Changbom Park's thesis, published in the *Monthly Notices of the Royal Astronomical Society* in 1990, proved that a cold dark matter inflationary model could produce structures as dramatic as the Great Wall of Geller and Huchra. It reproduced with remarkable fidelity the pattern of filaments and voids that they had observed. Gravity alone could produce these fantastic structures from the small initial fluctuations predicted by inflation. Exotic explanations, like nuclear explosions or cosmic string wakes, were not needed. One thing we knew about our simulations was that their galaxy clustering had a spongelike topology—clusters connected by high-density filaments and voids connected by tunnels. At this point we had only Geller and Huchra's 2D slice, not a fully 3D sample to work with, so we could not check their sample's 3D topology yet, but they were continuing to add to their slice, making it thicker and thicker, so eventually we would have a nice 3D volume to check for topology. For the time being, we knew that our simulations, which had a spongelike topology of large-scale structure, did produce the look of the Geller and Huchra slices perfectly. One did not require a froth of bubbles with a Swiss cheese topology to produce the Geller and Huchra slices.

The Great Attractor

Changbom Park and I continued to make larger and larger simulations and check them against new observations. In 1998, a group of seven astronomers (Donald Lynden-Bell, Sandra Faber, David Burstein, Roger Davies, Alan Dressler, R. Turlevich, and Gary Wegner) had discovered something they named the *Great Attractor*. This group of astronomers would become known as the "Seven Samurai." To measure distances of elliptical galaxies, they made use of an observed relation: the higher the *velocity dispersion* (or range of velocities) of the stars orbiting within an elliptical galaxy, the bigger in physical size the elliptical galaxy would be. The Seven Samurai could measure the velocity dispersion of the stars by measuring the broadening of the spectral lines in the galaxy. Different stars traveling at different speeds relative to us would create different redshifts for the spectral lines of those stars, smearing out, or broadening, the spectral lines seen in the spectrum of the galaxy as a whole. Once they knew the velocity dispersion of the stars in the galaxy, they could deduce its physical size. If they knew its physical size and then measured its angular size in the sky, they could deduce its actual distance. An object of a given physical size has a smaller angular size if viewed from a larger distance. If you know how big a "stop sign" really is, its angular size as you look out your windshield will tell you how far away it is. In this way the team could estimate the distances to many elliptical galaxies. If the distance to the galaxy differed from that deduced via Hubble's law from its redshift, it meant that the galaxy had a *peculiar*, or individual, motion over and above the general Hubble expansion. Our own galaxy, for example, has a *peculiar velocity* of about 550 kilometers/second relative to the cosmic microwave background. These peculiar velocities are, in the standard Big Bang inflationary cosmological model, produced by the extra gravitational attraction of nearby regions of above-average density (versus the lessened gravitational attraction of other nearby regions of below-average density). The Seven Samurai found that galaxies showed large peculiar velocities, all pointing to a Great Attractor located at a distance from us of about 200 million light-years in the constellation of Centaurus. They calculated that this must be due to an excess mass in this region of 5×10^{16} solar masses. They noted a concentration of galaxies seen in this region that was 20 times larger than the nearby Virgo Cluster (about 50 million light-years from us). The appearance of a Great Attractor was surprising. If the voids in the Geller and Huchra slices were produced by explosions, one might expect instead to find "Great Repulsors"—regions in the centers of voids from which galaxies were fleeing. Instead, when velocity flows in the universe were investigated, the first thing to appear was a Great Attractor.

Changbom and I decided to look for peculiar velocity flows in our simulations. We measured the gravitational potential energy at all points in our large cubic simulation. The low point in the gravitational potential—the most gravitationally bound spot in the entire simulation—should be a place toward which galaxies migrate. We knew the peculiar velocities of all the galaxies in our simulation, and so we put a slice through the most gravitationally bound spot and put peculiar velocity vector arrows on each galaxy showing where it was headed. We found a dramatic thicket of long arrows pointing right at the most gravitationally bound spot in the simulation—a Great Attractor! But why were there no Great Repulsors? We looked for the spot in the simulation that was highest in gravitational potential energy, the least gravitationally bound spot in the entire simulation. This should be a place from which galaxies were fleeing. But when we looked at a slice through the simulation that included this spot, we saw no thicket of long arrows headed away from it: no Great Repulsor. Why? It was simply because the least bound spot was located in the middle of a giant void, with no galaxies nearby to show the motion! It was, therefore, the lack of tracer galaxies nearby that made it unlikely for astronomers to discover a Great Repulsor. But a Great Attractor was demonstrably easy to find. Thus, the computer simulations had explained another, seemingly contradictory, discovery.

The velocity flows around the Great Attractor have now been mapped in an elegant way by R. Brent Tully, Helene Courtois, Yehuda Hoffman, and Daniel Pomarède (2014). At each position in space the peculiar, or individual, velocities of individual galaxies can be plotted as arrows using a surveying technique developed by Ed Bertschinger and Avishai Dekel (1989).[1] Tully and his colleagues then connect up the arrows to draw flow lines showing the direction of motion. These velocity flows are due to the gravitational attraction of neighboring groups and clusters of galaxies.

We may understand this by thinking about rivers and their watersheds on Earth. If you let a drop of water fall on the ground in a natural terrain, it will roll downhill, pulled by gravity. If you keep following its path, you will flow into a river and eventually arrive at the sea. The set of points where water can fall and eventually find its way into the sea at the mouth of the Mississippi river, for example, is called the watershed of the Mississippi. It is the total area drained by the Mississippi River. Tully and his colleagues have mapped the peculiar velocity flow lines of galaxies in our neighborhood. They show a giant watershed of flow lines leading to the Great Attractor, as if they were rivers (see Color Plate 5). We are part of this watershed. They have named this region the Laniakea Supercluster. (Laniakea means "immeasurable heaven" in Hawaiian.) We are at the outer edge of this region, which is 510 million light-years across.

We have a new addition to our cosmic address: Earth, Solar System, Milky Way, Local Group, Virgo Supercluster, Laniakea Supercluster. The Virgo Supercluster is just our nearby branch of the Laniakea Supercluster. Does this mean that our galaxy will eventually fall into the

Great Attractor? No, the individual velocity of our galaxy moving us in its direction is small relative to the average velocity of separation we are experiencing relative to it, due to the overall expansion of universe. Moreover, the expansion of the universe is accelerating, so the Great Attractor is actually fleeing from us ever more rapidly. But the Laniakea Supercluster marks the extent of the gravitational influence of the Great Attractor.

The Coma cluster is part of a different, neighboring watershed called the Coma Supercluster. Tully and colleagues identified two other independent watersheds: the Shapley Supercluster and the Perseus-Pisces Supercluster. We can divide the volume of space into these gravitational watersheds. This presents a new challenge for N-body computer simulations. We can use the same velocity flow techniques to map out watersheds in the N-body simulations and see if they look just like the ones we find in the universe. That will be an interesting area for future research. Given our ability to find Great Attractors in the simulations so easily, I expect N-body simulations to pass this test as well.

Rodlike Simulation

In 1990, Broadhurst and colleagues did two deep, pencil-beam surveys pointed in opposite directions in the sky to produce a rod-shaped survey with a length of about 10 billion light-years centered on Earth. They found a succession of walls in the survey that appeared to be approximately equally spaced, about 600 million light-years apart. That did not seem to fit in well with the idea of random initial conditions produced by inflation. Changbom and I immediately did a long cosmological rodlike simulation. It simulated a volume of 12 billion light-years × 490 million light-years × 490 million light-years (see Figure 8.2).

In this rod-shaped volume we also discovered a series of walls, and in simulating 12 narrow pencil-beam surveys, we found one survey that was even more nearly periodic in appearance than the survey by Broadhurst and colleagues. Thus, the periodic appearance of walls in the observational data was not statistically abnormal at the 95% confidence level in a cold dark matter inflationary cosmology. Random initial conditions with cold dark matter and inflation naturally produced a

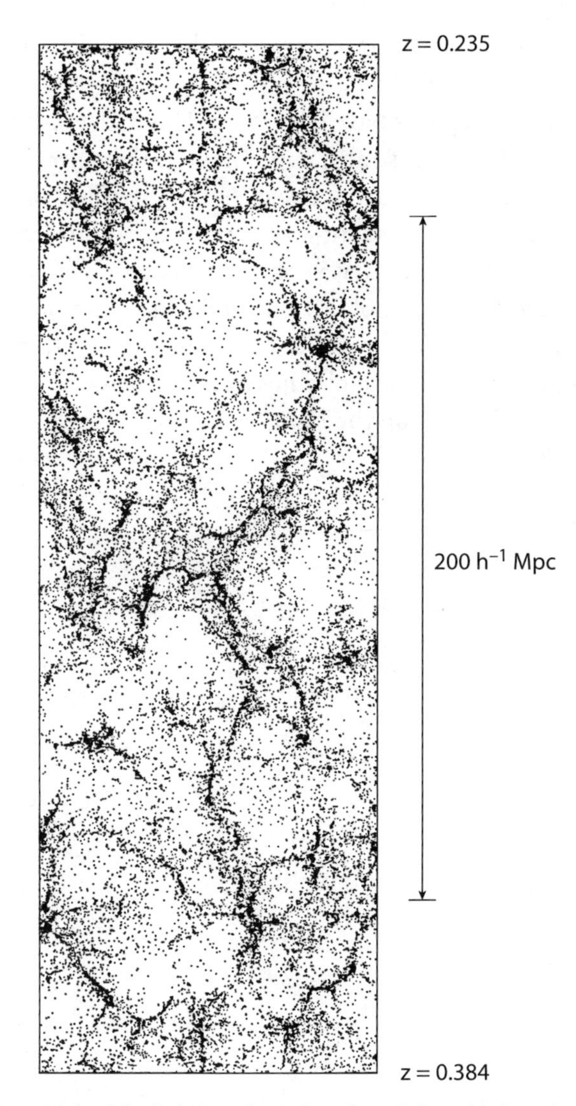

Figure 8.2. One-eighth of the Park-Gott simulation of a rod-shaped volume in the universe. Galaxies at the top are at a redshift of 0.235; galaxies at the bottom are at a redshift of 0.384. Scanning from top to bottom, one is looking further out in the universe and further back in time. The double arrows indicate a length of 973 million light-years. This section of the rod-shaped volume is 1.46 billion light-years tall, 487 million light-years wide, and 97 million light-years thick. One sees giant voids and great walls. (Credit: Changbom Park and J. Richard Gott, III, *Monthly Notices of the Royal Astronomical Society*, 249: 288, 1991)

sequence of walls in pencil-beam surveys that were just as dramatic as those found observationally.

Overall, the simulations have been remarkably successful. Thin slices of the universe show voids most prominently; the same holds true for simulations. Thicker slices show Great Walls—likewise for simulations. Velocity flows in the universe show Great Attractors, just as in the simulations. Deep pencil-beam surveys show a succession of Great Walls, as do the simulations. In cosmology, the larger the observational samples have become, and the larger the simulation volumes have become to match them, the more spectacular has been the agreement between the two. This is a sign we are on the right track.

Chapter 9

Measuring the Cosmic Web—the Sloan Great Wall

After our original papers on the topology of large-scale structure came out, we were anxious to apply the method to larger data sets. In the 1986–87 academic year, I took a sabbatical at the University of Virginia and worked with Trinh Thuan, John Miller, Stephen Schneider, David Weinberg, Charles Gammie, Kevin Polk, Michael Vogeley, Scott Jefferey, Suketu Bhavsar, Adrian Melott, Riccardo Giovanelli, Martha Haynes, Brent Tully, and Andrew Hamilton to analyze a large number of observational data sets. Surprisingly, I had formed a group of 15 people! This is as close to "big science" as I ever got! I am a theorist and usually work alone or with one or two colleagues, but this study analyzing many observational data sets required bringing together a lot of talented people. This is usually how it works: someone has an idea, and people come together to work on it. As science has become more complex, science collaborations have become larger. The WMAP satellite team that measured the cosmic microwave background grew to 21 members. The Sloan Digital Sky Survey, which I also work on (and will describe shortly), involves about 180 people—but I did not form that group. The Planck Satellite Collaboration reported new measurements of the cosmic microwave background in 2013 with a paper having 277 authors. (Of course, the really BIG science is over in physics, where the ATLAS Collaboration looking for the Higgs boson had a paper with an author list beginning *Aad*, *Abbott*, *Abdallah*, *Abdel*, . . . , and 3,062 more coauthors!) Today, the challenge for young astronomers is to figure out how to fit into the ever-larger collaborations. How does one get to lead a project? Should one participate in an already-large collaboration and work to advance to a position of leadership, or should one strike out on one's own and try to have ideas that will draw others in? In a highly competitive environment, it is always helpful to have an army at your back.

In this case, I was able to bring together a remarkable group of people to study topology. Trinh Thuan, from the University of Virginia, was my old friend from graduate school days at Princeton. At Virginia, Thuan and I studied of the distribution of dwarf and low-surface-brightness galaxies. These galaxies were too faint to have their redshifts measured by Geller and Huchra, but their positions in the sky were already listed in other catalogs. With Stephen Schneider, Thuan and I measured the redshifts of these galaxies using a radio telescope by observing the 21-centimeter wavelength line of atomic hydrogen. We discovered that these dwarf and low-surface-brightness galaxies fell on *exactly* the same structures that de Lapparent, Geller, and Huchra had found. They did not fill in the empty voids.

As for the rest of the topology group, Weinberg, Melott, and Hamilton were already on board. Michael Vogeley was originally an undergraduate at Princeton and is now a professor at Drexel; he and I have continued to work together for many years on topology. Charles Gammie and I worked on a sample of Abell galaxy clusters. Giovanelli and Haynes brought with them their sample of more than 4,000 galaxy redshifts. Tully would bring his catalog of nearby galaxies. He was already a master of cosmic cartography. With Richard Fisher he produced a wonderful atlas of nearby galaxies—it had the flavor of an old-time atlas, with page after page of charts showing their positions. Recently, he has collaborated on a 3D computer movie bringing this atlas into the twenty-first century, as well as mapping the Laniakea Supercluster.

One of the best things about having a collaboration of large but still manageable size is that the collaboration meetings are both fun and personal. We organized a meeting at Melott's home base, the University of Kansas, in April 1988. As I arrived in the hotel for the meeting, I saw all the happy faces of my friends. I was genuinely touched. Here was something we all were glad to get the chance to work on. I brought with me a bunch of red-blue 3D glasses with white cardboard frames, like those in Figure 6.14. I had red-blue slides of some of our spongelike simulations to show. Afterward, we signed each other's 3D glasses to take home as souvenirs. At the meeting I shared news I had heard about plans to build a 3.5-meter telescope on Kitt Peak outside Tucson, Arizona, for use as a survey instrument to take the redshifts of 100,000 galaxies. This survey, if completed, would be invaluable to our topology work. The number of topological features we could measure was directly proportional to the number of galaxies in the survey. We were used to seeing surveys containing a few thousand galaxies, so this would be a giant increase. Allotting telescope time always involves competition, but an additional tension exists between devoting a large telescope to individual observers who would apply for time for their own projects and devoting the telescope to a systematic survey that, when finished, would be useful to everyone. Geller and Huchra had shown the benefit of a large survey for understanding the universe. I hoped our meeting would provide some support for the large redshift-survey idea, which would benefit many other projects as well as our own.

Then Richard Kron from the University of Chicago got up to say that he had been thinking about building a special dedicated telescope to do a survey that would measure a million galaxy redshifts. A million! We were amazed and very excited. This would be a dream survey for our topology research. We gave this our hearty endorsement. Ultimately, the people at Kitt Peak never built the survey instrument, but Richard Kron took his idea for a survey telescope to Jim Gunn at Princeton. Gunn decided to expand the idea. The telescope would be both a camera and a redshift-measuring machine. It would produce a digital sky map using a massive CCD digital camera; then, after selecting target galaxies from the sky map, it would use fiber optic cables plugged into the focal plane of the telescope to take hundreds of galaxy spectra at a time. Gunn, a master telescope and camera designer (having been a principal designer of the Wide Field and Planetary camera for the Hubble Space Telescope), devised the whole system—to be placed on a 2.4-meter-diameter telescope at Apache Point, New Mexico. This would become the Sloan Digital Sky Survey, a joint project of the University of Chicago, the Fermi National Accelerator Laboratory, the Institute for Advanced Study, the Japan Participation Group, the Johns Hopkins University, Princeton

University, the U.S. Naval Observatory, and the University of Washington. Financial support was provided by the Alfred P. Sloan Foundation, the (U.S.) National Science Foundation, the Japan Ministry of Education, and the participating institutions. It has been amazingly successful.

Our topology group analyzed all the available samples and presented our results in an omnibus paper in 1989. We found that when we smoothed the data and produced density contours, the median density contour *always* had a spongelike topology. As an example, the Giovanelli and Haynes sample is shown in Figure 9.1.

This sample of 491 galaxies extends to a distance of 574 million light-years. We applied a smoothing length of 41 million light-years; that is, we blurred each of the galaxies over a scale of 41 million light-years to create a smooth, undulating density field that captures the clustering pattern. We then constructed density contour surfaces. We show here the 50% (median) density contour surface that divides the survey region into a high-density half and a low-density half by volume. On the left we

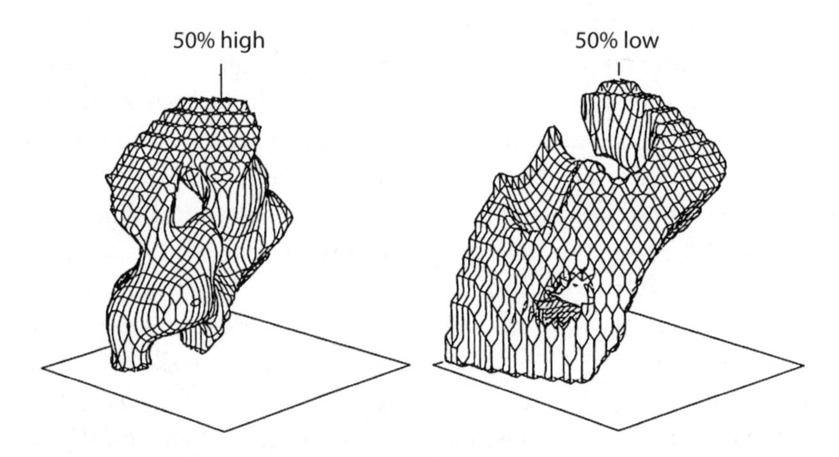

Figure 9.1. Spongelike topology in the Giovanelli and Haynes sample. The 50%-high-density regions appear at left, the 50%-low-density regions at right. Add the two regions together and you have the whole observational sample. The high- and low-density regions are spongelike and interlocking, with donut holes characteristic of a spongelike topology in agreement with Gaussian random-phase initial conditions. Earth is situated at the back corner of the white square (under each region). The Perseus-Pisces supercluster is contained within the linear structure at the bottom right of the 50%-high-density region. The curved outer edge of the observational sample is at a distance of 574 million light-years from Earth. The data have been smoothed with a smoothing length of 41 million light-years. (Credit: J. Richard Gott et al. *Astrophysical Journal*, 340: 625, 1989)

show the high-density half. Earth is in the back, at the back corner of the bottom square, behind the survey region. A big high-density filament passes though the survey region. This is the elongated Perseus-Pisces Supercluster. The low-density 50% is shown on the right. If you were to push the two regions together, they would add up to give the entire survey. You can see the curved spherical outer edge of the survey at a distance of 574 million light-years from Earth. The topology is clearly spongelike. The high- and low-density regions are interlocking, with donut holes apparent in both the high- and low-density regions. The genus curve for the Giovanelli and Haynes sample has the expected "W" shape as a function of density, agreeing with the theoretical Gaussian random-phase curve, within the errors.

Overall, we analyzed 12 different data sets of various sizes at various smoothing lengths. We could compare these genus curves with standard biased cold dark matter (CDM) simulations and with simulations in which galaxies were laid down on bubble walls (a Swiss cheese topology). The simulated bubbles had an average diameter of 156 million light-years, typical of the void sizes seen in the Geller and Huchra slice. We could then ask for the odds that the observational genus curves had been drawn at random from the CDM simulations as opposed to the bubble simulations. We found a probability of 98.8% that the observations resembled the CDM simulations as opposed to the bubble simulations. This was an impressive victory for the CDM model and for the spongelike topology expected from inflation.

The IRAS Sample

In 1992, an independent group, headed by Ben Moore from the University of Durham, analyzed the topology of large-scale structure using a sample of galaxies detected in the infrared using the Infrared Astronomical Satellite (IRAS). Other members of the group were Carlos Frenk, David Weinberg, Will Saunders, Andy Lawrence, Richard Ellis, Nick Kaiser, George Efstathiou, and Michael Rowan-Robertson. It was quite a distinguished group, full of leaders in the field. Observing in the infrared penetrates through the dust in our Milky Way, allowing astronomers to survey the entire sky without having to avoid the galactic plane, as

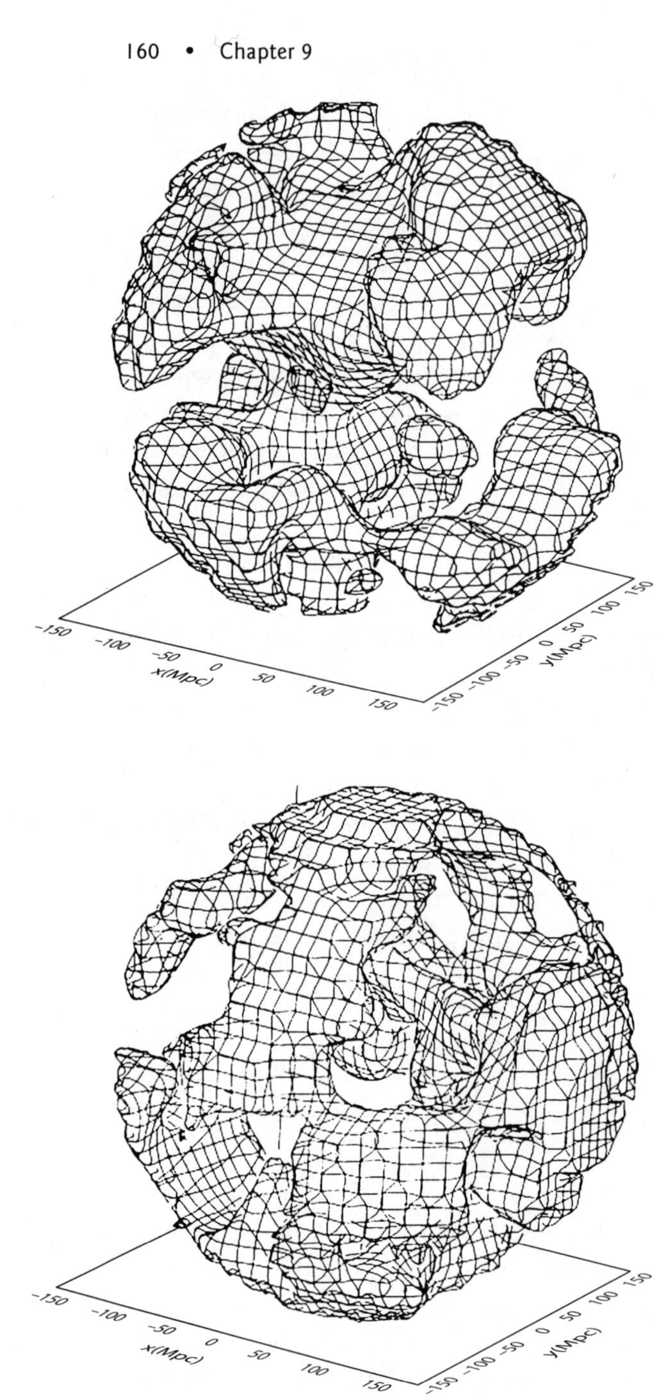

Figure 9.2. The topology of the Moore and colleagues' IRAS sample is shown: high-density 50% (*top*) and low-density 50% (*bottom*). Add the top and bottom halves to get the entire spherical sample centered on Earth with a radius of 487 million light-years. The smoothing length is 52 million light-years. The galactic plane of the Milky Way is horizontal. The upper hemisphere is the North Galactic Hemisphere, and the bottom hemisphere is the South Galactic Hemisphere. (Credit: Ben Moore et al., *Monthly Notices of the Royal Astronomical Society*, 256: 477, 1992)

one must when observing galaxies in the visible spectrum. Figure 9.2 shows the IRAS sample with the high-density 50% at the top and the low-density 50% at the bottom. The high-density regions form one connected piece, as do the low-density regions. It's a spongelike topology.

If one plots the high-density 10%, the contour breaks up into individual clusters, as expected, as seen in Figure 9.3. This survey went out to a smaller distance.

Similar plots of low-density regions showed famous voids previously charted by Brent Tully. Given the symmetric, W-shaped, Gaussian random-phase genus curves they were seeing, the IRAS group could rule out a bubble topology having approximately 486-million-light-year-diameter cells with a probability of 99.9999%. From the number of structures being found as a function of the smoothing length, they could estimate the slope of the power spectrum, which implied *more* power at large scales than would occur with a random (Poisson) sprinkling of points in the initial conditions. This was the same result Ed Turner and I had obtained by analyzing the frequency distribution of galaxy clusters of different sizes. All this was quite encouraging. It is always important to have independent groups verify your findings. This IRAS sample extended far

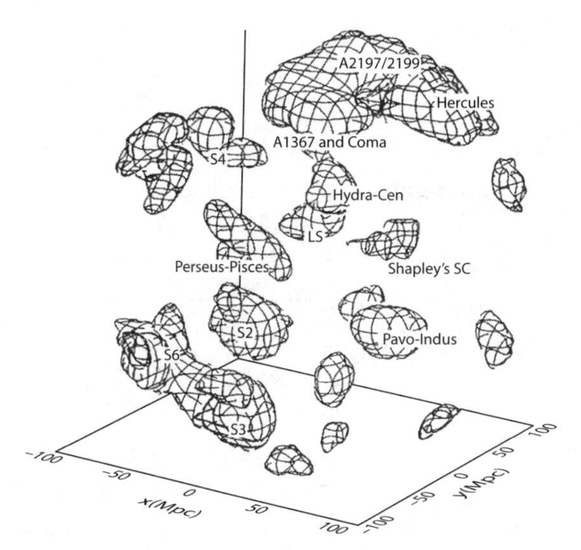

Figure 9.3. The topology of the high-density 10% of the IRAS sample shows isolated clusters and superclusters as expected for Gaussian random-phase initial conditions. The big clump at the top includes the Coma Cluster and the densest part of the Great Wall of Geller and Huchra. The Perseus-Pisces supercluster, which formed the main part of the high-density region in Figure 9.1, appears here as well but at a smaller scale and from a different angle. (Credit: Ben Moore et al., *Monthly Notices of the Royal Astronomical Society*, 256: 477, 1992)

enough to include some of the Great Wall region, and we were pleased that it showed the spongelike topology we expected from our theory. Another study of the topology of a sample of IRAS galaxies by Canavezes and colleagues in 1998 also found spongelike structure, lending further support to the random-phase nature of the initial conditions.

Vogeley and the Topology of the Great Wall Region

Meanwhile, by 1993 Geller and Huchra at Harvard had completed their 3D volume sample. They allowed us to cooperate with them on its topological analysis. Michael Vogeley, my former Princeton undergraduate student, had by this time moved on to Harvard as a graduate student and was leading the study, with participation by Changbom Park, Geller, Huchra, and me. Figure 9.4 shows the high-density 50% of the survey using a small smoothing length (on the left) and a larger smoothing length (on the right).

The orientation of this figure is the same as in the previous two figures. The plane of the Milky Way galaxy is horizontal. The North Galactic Hemisphere is at the top, and the South Galactic Hemisphere is at the bottom. The expanded CfA survey, shown on the left, included a really thick fan-shaped region in the north, and another substantial cone-shaped region in the south. Earth is at the center of the figure. The bar at the bottom has a length of 243 million light-years. The outer radius of the survey from Earth is 608 million light-years. The smoothing length (in the left-hand picture) is 28 million light-years. The Great Wall snakes from left to right across the upper fan-shaped survey region, with a length of 758 million light-years. The high-density 50% region is in one connected piece and has a spongelike topology. The voids on the near and far sides of the Great Wall are connected by low-density passages. You can swim from the void on the near side to the void on the far side through low-density regions by simply going around the Great Wall. It is a filament. To the right we show the same 50% high-density regions but with a smoothing length of 56 million light-years. This gives a low-resolution look at the same survey. It is smoother and, therefore, shows less detail. The Great Wall clearly appears as a thick filament. A donut hole in the high-density region can be seen above it. The higher

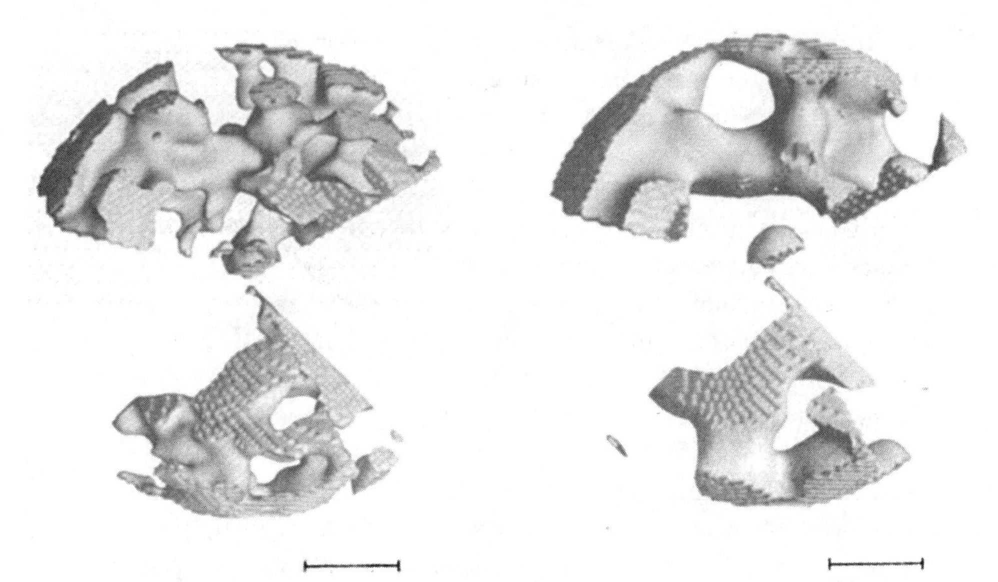

Figure 9.4. The complete 3D Center for Astrophysics (CfA) survey sample's 50%-high-density regions are shown at two smoothing lengths: 28 million light-years (*left*) and 56 million light-years (*right*). The Great Wall extends across the upper fan-shaped survey region, with a length of 758 million light-years. The 50%-high-density region is in one connected piece with a spongelike topology. The Great Wall is a filament. (Credit: Michael S. Vogeley, Changbom Park, et al. *Astrophysical Journal*, 420: 525, 1994)

resolution picture at the left reveals a finer, more intricate web of minor filaments that the low-resolution picture leaves out, but in both pictures the high- and low-density regions are interlocking and spongelike. A close-up (upside down) high-resolution picture of the Southern extension of the survey is shown in Color Plate 6.

In this color picture of the Southern CfA Survey, we see the 50% high-density region at the top, where Michael Vogeley has placed a (simulated) red-light source in the middle of the large void in the middle of the survey. You can see the red light shining out into other void regions through tunnels between the voids. At the bottom you can see the 50% low-density regions. If added to the high-density regions, they would produce the entire survey region. The large void in the center can now be seen as a solid blob with extensions (low-density tunnels) connecting it to other void regions. This is again a spongelike topology: the high- and low-density regions are multiply connected and interlocking.

This CfA survey marked an acid test of the spongelike topology, because initially, after the first 2D slice of the region was shown, it looked to Geller and Huchra like a froth of bubbles with a Swiss cheese topology. It appeared that the voids might be surrounded by dense walls on all sides, creating isolated chambers or cells. But when the full 3D structure could be quantitatively measured, it showed a spongelike topology. Voids were connected to other voids by low-density tunnels, and great clusters were connected to each other by dense filaments of galaxies. This was the interlocking spongelike configuration expected from cold dark matter and inflation, a geometry that could be formed from random quantum fluctuations in the early universe. The Great Wall was simply a big filament connecting clusters, part of the overall spongelike structure we now call the *cosmic web*.

Once a filament formed, gravity would draw it together, making it even narrower. This is a nonlinear effect. Just as a cluster collapses due to gravity, a filament will collapse in the directions perpendicular to its length, with material crashing together to form a very narrow filament. As Zeldovich had imagined, the waves in the original universe would break, like waves in the ocean, and form caustics as the fluctuations became nonlinear fractional density enhancements greater than one. A structure like the Great Wall would form in this way. As the universe expanded, the cosmic web itself would stretch, and the filaments connecting clusters would become longer and narrower as gravity squeezed them and the expansion lengthened them.

Origin of the Name *Cosmic Web*

Where did the name *cosmic web* come from? The first paper to use it in either its title or abstract was the 1995 *arXiv* preprint posted online by Richard Bond, Lev Kofman, and Dmitry Pogosyan. The title of their paper was "How Filaments Are Woven into the Cosmic Web" (1995). Its abstract stated the following:

> Observations indicate galaxies are distributed in a filament-dominated weblike structure. Numerical experiments at high and low redshift of viable structure formation theories also show filament-dominance. We present a

> simple quantitative explanation of why this is so, showing that the final-state web is actually present in embryonic form in the overdensity pattern of the initial fluctuations, with nonlinear dynamics just sharpening the image.

This agrees, of course, with what we had been saying all along, that the sponge was present in the initial conditions and is the same sponge we see today—just enhanced in contrast by gravity. Bond and his colleagues started off their paper with a picture of a computer simulation (showing the present epoch) from Anatoly Klypin using the cold dark matter model. It looked decidedly filamentary (see Figure 9.5).

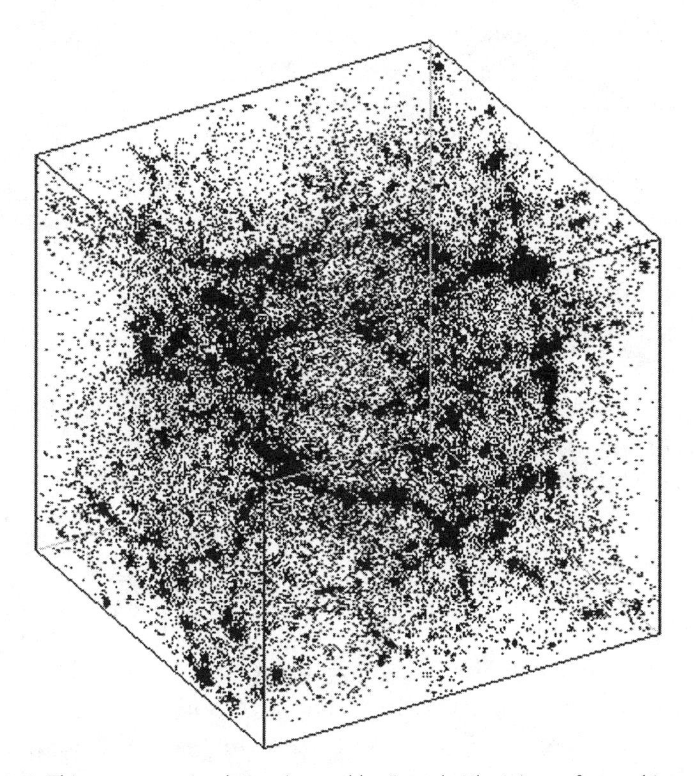

Figure 9.5. This computer simulation (created by Anatoly Klypin) was featured in a paper by Bond and colleagues (1995), which coined the term *cosmic web*. This simulation shows the spongelike topology that results from inflationary initial conditions, with narrow filaments connecting clusters in a cosmic web. Voids are connected by tunnels. (Credit: Courtesy of Anatoly Klypin, as appearing in J. R. Bond, L. Kofman, and D. Pogosyan, arXiv:astro-ph/9512141v1, 1995)

Bond and his colleagues were interested in asking why the filaments were so dominant. They analyzed the initial conditions in the computer simulations, looking for patterns in velocity flows. A filament in the initial conditions would produce, via gravitational attraction, peculiar velocity flows drawing particles inward in two dimensions: radially inward toward the linear filament. Imagine the filament as a vertical rod: it would draw particles in toward itself in two horizontal directions by gravitational attraction—like moths being attracted toward a long fluorescent tube. An initial pancake, on the other hand, would produce peculiar velocity flows that were directed inward in one dimension—perpendicular to the pancake. If you imagine a horizontal pancake, it will produce infall velocities just along the single vertical dimension as particles fall toward the pancake from above and below it. The authors found a predominance of initial flows in the computer simulation in two dimensions, as would occur for filaments in the initial conditions. That indicated a predominance of filament-shaped density fluctuations in the initial conditions, as opposed to pancakes; as the universe evolved, nonlinear gravitational effects would simply have made the filaments thinner. Their pictures showed that in the final conditions, the clusters were connected by thin filaments of galaxies. These were, of course, computer simulations based on Gaussian random-phase initial conditions that were meant to explain the structure *already* seen in the observations by our group and others.

Where did the word *web* come from? Bond and colleagues started their 1995 paper by showing a computer simulation by Klypin using cold dark matter. In the original 1983 computer-simulation paper. in which Klypin and Shandarin had written about the Zeldovich honeycomb model, they had used the word *web*. Here is their sentence (which I have quoted before): "The regions of high density seem to form a single three-dimensional web structure." They then noted, "However it is not clear from our simulation whether honeycomb structure arises or not." With hindsight it is clear that the initial conditions they were using were Gaussian random phase, so the topology in the initial conditions had to be spongelike. They had initial conditions in which fluctuations on small scales were damped out. Melott, Dickinson, and I had also examined such models in our initial topology paper (see Figure 6.7), along with CDM simulations, and found that they were *all* spongelike in the initial conditions and that the initial sponges were replicated almost exactly, in enhanced contrast, in the final conditions. Klypin and Shandarin's computer simulations did not form a honeycomb, as they and Zeldovich would have hoped, but instead formed a sponge.

The term *cosmic web* captures both the topology of the large-scale structure and the thinness of the filaments, reminiscent of a spider's web. That name has stuck. The Bond and colleagues' *arXiv* paper would eventually be published in 1996 in *Nature*, with the slightly revised title "How Filaments of Galaxies Are Woven into the Cosmic Web."

The Sloan Great Wall

If we could see a giant filament in the cosmic web as close to us as the Great Wall, then if we looked deeper in the universe, we might find others. Indeed, our topology group did a computer simulation at the turn of the millennium to show what structures might be found in Jim Gunn's upcoming Sloan Digital Sky Survey. Wes Colley led the project, with Andreas Berlind joining Changbom Park, David Weinberg, and me. We had 54 million CDM particles in a cubic volume 2.9 billion light-years on a side to simulate the million galaxies that might be included in the Sloan Digital Sky Survey 3D map. (This was one of the first simulations that also included dark energy—a uniform component of the universe discovered in 1997, which we will discuss in Chapter 11. All subsequent simulations include dark energy.) Our simulation produced a spongelike topology, whose median density contour was, of course, spongelike. Nonlinear gravitational effects produced slightly more isolated clusters in the 7% high-density sample than isolated voids in the 7% low-density sample in the final conditions. In other words, the 7% low-density volume is divided into fewer and bigger voids, whereas the 7% high-density volume is divided up into smaller, more numerous clusters. Gravity causes clusters to contract, making them smaller, and being smaller, it takes more of them to make up 7% of the volume. By contrast, voids expand, becoming bigger, so fewer of them are needed to make up the low 7% of the volume. As the fluctuations became of order 1 or more—that is, the density in the highest density regions became denser than 1 + 1, or 2, times the average density in the universe—the fluctuations begin to grow at a faster than linear rate compared to the expansion of the universe. This is called *nonlinear evolution*. It is what Jim Gunn and I were investigating when we followed the collapse of the Coma cluster. This kind of evolution has to be followed on a computer.

The result that clusters should slightly outnumber voids due to nonlinear effects was discovered in 1994 by Takahiko Matsubara, at the University of Tokyo, who solved the equations for mildly nonlinear gravitational evolution and applied them to topology. The genus curve retained its familiar W shape but was slightly distorted. Isolated clusters slightly outnumbered isolated voids. (This effect can be seen in the final and biased conditions in the computer simulations shown in Figure 6.10, top graph.) The central peak in the genus was also shifted slightly to the left—what we called a *meatball shift*—because of the prominence of big, high-density features like the Great Wall. Such effects could be accentuated by biased galaxy formation—the fact that galaxies are more likely to form in high-density regions. We had noted such effects already in our omnibus observational paper of 1989; the Sloan Digital Sky Survey would offer a chance to see them in greater detail.

Our computer simulation even found what we called a "great wall complex" extending for a distance of 1.9 billion light-years across the survey. Thus, we predicted that the Sloan Digital Sky Survey would find great walls significantly larger than the Great Wall of Geller and Huchra. This prediction would be vindicated in spades.

In 2003, when the first Sloan data finally came in, Mario Jurić and I set ourselves the task of mapping its structure in a visually compelling way. The first data release was complete for an equatorial slice 4° thick extending in two giant fans out from Earth's equator. One fan included the Northern Galactic region and the other, the South. The survey avoided the two regions where the Milky Way galactic plane crossed Earth's equatorial plane. Figure 9.6 shows two fans extending from Earth at the center. Each point represents a galaxy. Earth is at the center. We are looking straight down on the two slices from a viewpoint far above Earth's North Pole. Distances are plotted as comoving distances, showing the distances the galaxies will have at the present epoch (as defined in note 1 for Chapter 3).

Mario and I saw immediately that the Northern slice had an enormous wall. It had shown up earlier as one connected high-density region

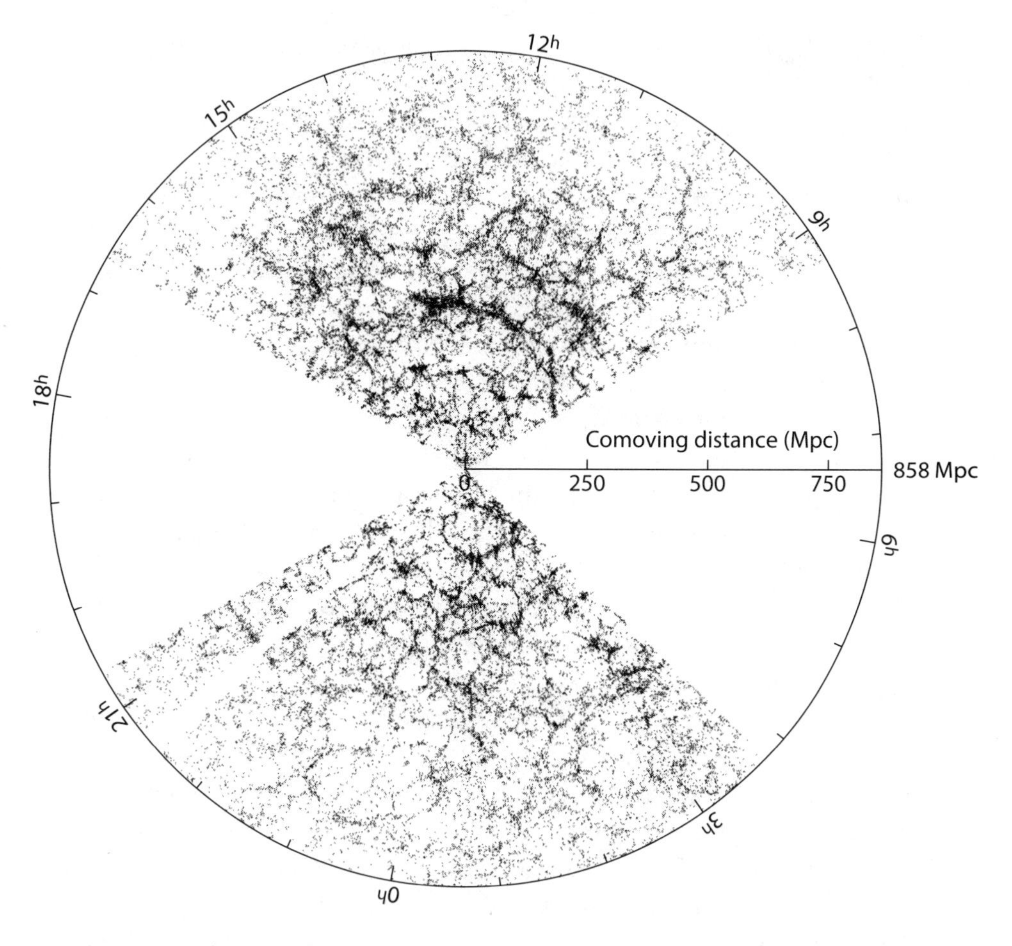

Figure 9.6. This equatorial slice of the Sloan Digital Sky Survey (SDSS) from Gott, Jurić, and colleagues, shows a 360° panorama looking out from Earth's equator. Earth is at the center. The points represent galaxies within 2° of the celestial equator (which passes overhead of Earth's equator). The two fans show the region covered by the SDSS; the two blank regions are close to the Milky Way's galactic plane and are, therefore, not surveyed. The radius of this picture is 2.80 billion light-years. Many voids and filaments are visible, the most prominent filament being the Sloan Great Wall in the top fan. (Credit: J. Richard Gott, Mario Jurić, et al., *Astrophysical Journal*, 624: 463, 2005)

in a topology study of a volume-limited sample of this slice (including only intrinsically bright galaxies), which I had done with Fiona Hoyle, Michael Vogeley, and other colleagues in 2002. I thought to name it the *Sloan Great Wall*, which would honor everyone in the survey. The Great

Wall of Geller and Huchra had by now become known as the CfA2 Wall, named after the Center for Astrophysics Survey 2 in which it was discovered. In Figure 9.6 you can also see many small voids, just like the ones found by Geller and Huchra. A particularly nice void is located near us in the southern fan (at bottom); it is punctuated by a couple of rich clusters, which appear as tiny daggers pointing at Earth due to their large internal velocities. The lacework of the cosmic web is apparent. The texture of this picture strongly resembles the texture of our large, rod-shaped, computer simulation using cold dark matter shown in Figure 8.2. In our paper we included a side-by-side comparison of the Sloan Great Wall and the Great Wall of Geller and Huchra (see Figure 9.7).

Mario and I measured the length of the Sloan Great Wall—it was 1.37 billion light-years long, 1.8 times as long as the Great Wall of Geller and Huchra. It was quite like the original Great Wall, only bigger! Like the Great Wall, the Sloan Great Wall is not gravitationally bound but is a coherent structure that will retain its identity as the universe expands. In the future, it will continue to stretch in length as the universe (and, therefore, the cosmic web) expands, while gravity will keep it as narrow as it is now or make it even narrower, as time goes on. The left hand (Eastern) end of the Sloan Great Wall is particularly spectacular, containing a number of rich clusters. Then it splits into two strands, like a multilane highway becoming a divided highway for a while, before rejoining to make one strand again at the western end. Lorne Hofstetter and I would make a spectacular portrait of the Sloan Great Wall, 60 inches wide, which included tiny pictures of each galaxy from the Sloan data. A small part of this grand view, a close-up of the Eastern end of the Sloan Great Wall, appears in Color Plate 7. To make this picture, we had to enlarge each galaxy by a factor of 50 relative to its true size in order to make each galaxy visible. Put another way, if the distances between the galaxies shown on in the picture were depicted properly, the picture would have to be 50 times larger.

The Sloan Great Wall was reported in *Science*, the *New York Times*, *New Scientist*, and *Scientific American*, among others. The Sloan Great Wall would eventually end up in the *Guinness Book of Records* 2006, as the "largest structure in the universe." This was a record the Great Wall of Geller and Huchra had held previously. There are many "largest" things in the Guinness book, but this is the largest of the largest. Mario

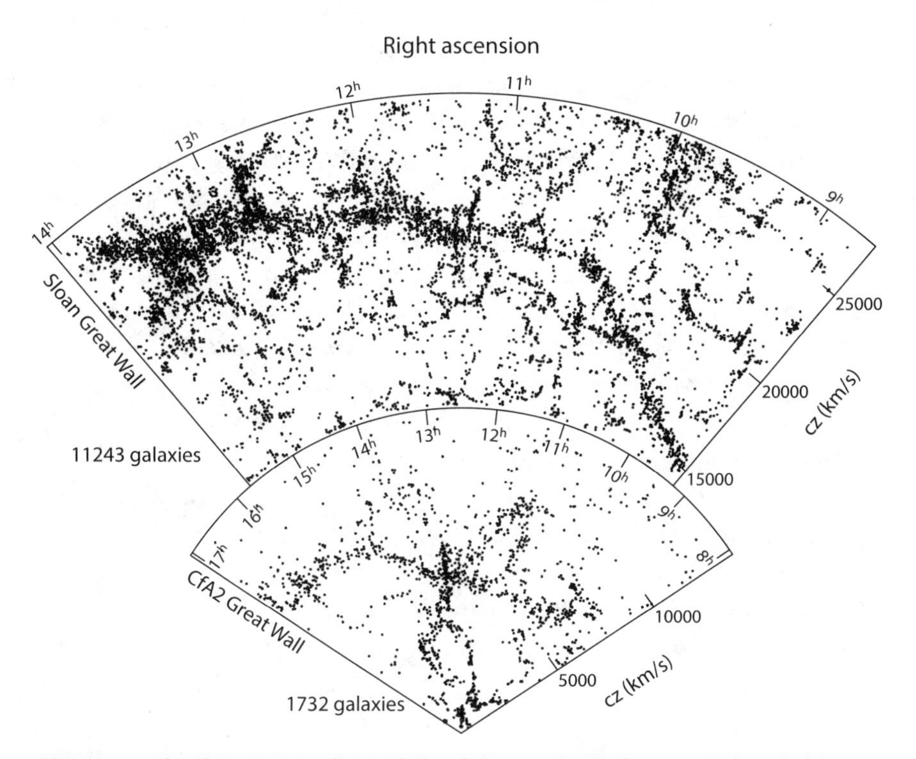

Figure 9.7. The Sloan Great Wall (1.37 billion light-years long) shown to scale with the CfA2 Great Wall of Geller and Huchra (758 million light-years long). Earth is at the bottom point of the fans. (Credit: J. Richard Gott, Mario Jurić, et al., *Astrophysical Journal*, 624: 463, 2005)

and I were both mentioned in the entry. (It held the *Guinness* record till 2015, when surpassed by a wall from a deeper survey.)

The length of the Sloan Great Wall is 1⁄10 the radius of the entire visible universe (everything we can see with our telescopes) out to the cosmic microwave background. To put this into perspective, if you made a scale model of the entire visible universe that was as big as Earth, the Sloan Great Wall would be 400 miles long. On this scale, our Milky Way would be about 150 feet across, and the Andromeda Galaxy would be of similar size, about 7⁄10 mile away. The distance between the Sun and Proxima Centauri, the nearest star, would be only 1⁄13 inch. The solar system would be microscopic on this scale.

To capture the entire universe, including all these scales, on one map, Mario and I presented a logarithmic map in our paper. From left to right, it represented a 360° panorama looking out from Earth's equator. The vertical coordinate measured distance from the center of Earth, with equally spaced tick marks vertically marking increases by factors of 10 away from Earth: 1 Earth radius, 10 Earth radii, 100 Earth radii, and so forth. The map was conformal, preserving shapes locally, so features like the Sloan Great Wall were neither squashed nor stretched vertically. As one moved up the map, from Earth's surface (a horizontal line at the bottom of the map), one first encountered artificial satellites orbiting Earth; then the Moon, Sun, and planets; then stars, galaxies, and quasars; and finally, the cosmic microwave background, the most distant thing we can see, at the top (see Color Plate 8, a color version of the top one-sixth of the map).

The *Los Angeles Times* called it "arguably the most mind-bending map to date." *New Scientist*, *Astronomy*, and the *New York Times* have reprinted copies of it. The most prominent features in the top section of the map shown in Color Plate 8 are the Whirlpool and Sombrero Galaxies, the Coma Cluster of Galaxies, the CfA Great Wall of Geller and Huchra and the Sloan Great Wall. Objects further up the map, further away from Earth, are shown at smaller scale. The Sloan Great Wall is 3 times further away than the CfA Great Wall and is, therefore, shown at ⅓ the scale; because of this scaling factor, although it is actually about twice as long, the Sloan Great Wall appears ⅔ the size of the CfA Great Wall on this map. The two vertical reddish bands show the regions covered by the SDSS survey. Each red dot is a galaxy or quasar. The blank bands were not covered by the survey. The Cosmic Microwave Background covers the entire sky, stretching 360° across the entire top of the map.

The Millennium-Run Computer Simulation

Perhaps the most beautiful computer simulation of what the cosmic web looks like was produced in 2005 by Volker Springel and his colleagues, who did what was also the largest *N*-body computer simulation of the time—the Millennium Run—using 10 billion particles to simulate the mass in a cubic volume that has expanded to a size of 2.43 billion light-years on a side at the present epoch. Color Plate 9 presents a slice through the simulation 73 million light-years thick. It shows beautiful filamentary structures connecting the clusters. Some have likened this picture to one of brain neurons interconnected by synapses. A rich cluster in the center, surrounded by strong filaments, can be seen.

In Color Plate 10, several fans are shown. The observations (shown in blue) are the CfA and Sloan Digital Sky surveys (at the top) and the 2dF survey (Colless et al. 2001), using the Anglo-Australian Telescope's 2°-wide, multiobject spectrograph (at the left). The 2dF Galaxy Redshift Survey (sometimes called 2dFGRS or 2dGRS) is a particularly interesting thick-slice survey. Studies matching the power spectrum of fluctuations in galaxy counts on different scales in the 2dF to the inflationary model allowed Percival and colleagues (2001) to estimate the quantity $\Omega_{\text{matter}} h = 0.2$, where $h = H_0/100$ kilometers/second/megaparsec. With the current best value of $h = 0.67$ from the Planck satellite team and the Sloan Digital Sky Survey, this leads to a value of $\Omega_{\text{matter}} = 0.3$. This estimate is based on the 3D locations of the galaxies in space. Separate studies of the magnitude of velocity flows in the 2dF survey led to a completely independent measurement of $\Omega_{\text{matter}} = 0.3$ (Peacock, Cole et al. 2001). One measurement is based on galaxy positions, the other on galaxy peculiar velocities. Both measurements of Ω_{matter} agree with each other and with the current estimate of $\Omega_{\text{matter}} = 0.308$ from the Planck Collaboration (2014) using the cosmic microwave background radiation. In red (at the bottom and right) are equivalent fans produced from the Millennium Run simulation, which found one great wall complex that was as long as the Sloan Great Wall but not as spectacular. The Millennium Run simulation used $\Omega_{\text{matter}} = 0.25$ and $h = 0.73$, close to the best estimates ($\Omega_{\text{matter}} = 0.308$ and $h = 0.67$) available today. The overall agreement between the Millennium Run simulation and the observations is impressive. We don't expect it to produce structures that are exactly the same, just ones that are similar in their statistical properties. In this it succeeds splendidly. The overall look of the slices compares well.

Juhan Kim and the Horizon Run Simulations

The Sloan Great Wall is so dramatic that one might wonder if it is consistent with the standard CDM inflationary cosmology. To answer this question, Juhan Kim, in collaboration with Changbom Park and other members of our group, did a series of even larger computer simulations called the Horizon Run simulations. The 2011 Horizon Run 3 used a record-breaking 374 billion particles to simulate a cubic volume 53 billion light-years on a side at the present epoch. It is 45,000 times the volume of Park's original simulation! The particles in the simulation represent cold dark matter. The dynamics of these particles due to the action of gravity are accurately tracked down to a resolution of 730,000 light-years, less than a third the distance from us to the Andromeda Galaxy. Galaxies are identified with cold dark matter halos that are gravitationally bound and dense enough not to be disrupted by tidal gravitational forces from neighboring halos. These are places where we expect normal atoms to congregate, dissipate energy, and form galaxies. From this large sample volume, we can produce 27 independent mock Sloan Digital Sky Survey catalogues. This simulation was performed on the Korean Supercomputer and is large enough to capture the true large-scale structure of the universe and give enough samples to provide good statistics for rare features like the Sloan Great Wall.

We used a standard linking algorithm to identify superclusters. Galaxies are considered "friends" if they are separated by less than a certain linking distance. A supercluster is defined as a set of "friends of friends" of a given galaxy. If the linking distance is too small, all galaxies will be singles and there will be no superclusters. Make the linking distance longer, and superclusters will start to form as galaxies link up. But if the linkage becomes too large, eventually the number of superclusters begins to drop as they link up to form the one spongelike high-density region, which is all in one connected piece. We chose our linkage length so that the *maximum* number of superclusters would be selected. (This picks more, and smaller, superclusters than might be found if a larger linkage length were used. But it is a well-defined procedure.) This algorithm was then applied in identical fashion to the SDSS observations and the mock SDSS surveys from the simulations. Then the richest supercluster was selected (the one with the most members). In the observations, this coincided with the rich eastern end of the Sloan Great Wall. It could then be compared with the median richest supercluster from the simulation's mock surveys picked the same way. A similar algorithm was used to find the largest volume void complexes—the largest connected regions of especially low density. The largest void complex in the observations was likewise compared with the median largest void complex from the simulations. These comparisons are shown (at the same scale) in Color Plate 11.

The Sloan Great Wall is quite typical of the most dramatic structures we would expect in a volume as large as the Sloan survey. The richest structure we show from the simulations is the median, which means that half the simulated Sloan surveys contained a richest structure richer than this one. And the largest void complex seen in the observations—a large cavernous volume of connected low-density chambers—is also similar to the largest void complexes typically found in the simulations. The simulations show a range for the largest expected supercluster and the largest expected void complex in a Sloan-sized survey. A detailed study showed that sizes of the Sloan Great Wall and the largest observed void complex fit within the 95% confidence limits estimated from the simulations.

Topology of Large-Scale Structure in the Sloan Digital Sky Survey

Once we had surveyed the equatorial slice of the Sloan Digital Sky Survey, the next task was to measure the Sloan 3D topology, a project that Clay Hambrick, Michael Vogeley, and I did, along with our colleagues from the topology group and colleagues in the SDSS.

Figure 9.8 shows the progressively larger samples we have studied, from the tiny 1986 original CfA sample, to the 1994 Geller and Huchra sample, to the 2006 Sloan Digital Sky Survey (SDSS) sample. For the 2006 SDSS sample at right, Earth is in the background, and the two lung-shaped regions are the partially completed regions of the Sloan survey in the foreground. The high-density 50% of the survey is shown as solid. One can easily that see the high-density regions are in a multiply connected spongelike piece. The Sloan Great Wall goes across the bottom lunglike survey region. The low- and high-density regions are completely interpenetrating. At this point in time (2006), the whole survey included about 400,000 galaxies. As we have looked at larger and larger samples, the topology has consistently remained spongelike.

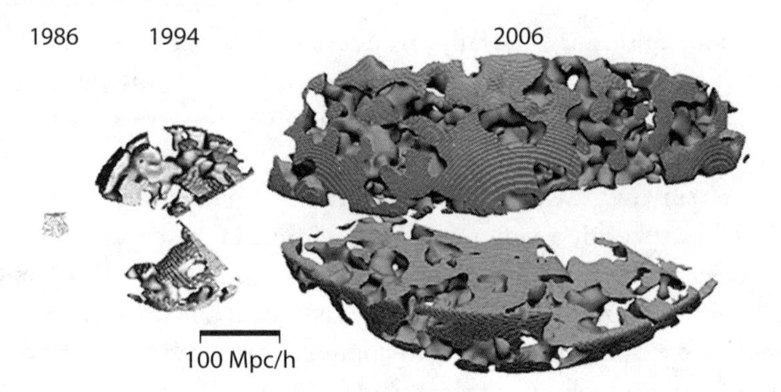

Figure 9.8. The spongelike topology of the 50%-high-density regions (solid areas) is illustrated by three examples (shown at the same scale) of the ever-larger topology surveys that have been done with time. The black line has a length of 100 Mpc/h, or 487 million light-years. At the left is our first small CfA cubical survey from 1986. In the middle is the Vogeley, Park, and colleagues (1994) topology study of the final 3D CfA2 sample, including the Great Wall at the top. At right is the Gott, Hambrick, Vogeley, and colleagues (2008) Sloan Digital Sky Survey sample; the Sloan Great Wall stretches across the top of the lower lunglike survey region. All are spongelike. (Credit: Adapted from: J. Richard Gott, D. Clay Hambrick, et al., *Astrophysical Journal*, 675: 16, 2008)

The results we were getting from the Sloan Survey using galaxies to trace large-scale structures were supported by an independent survey of cold dark matter made with the Hubble Space Telescope using gravitational lensing (Richard Massey et al. 2007). By measuring the distortion in the shapes of background galaxies, the total matter distribution, which is dominated by cold dark matter, can be mapped, as was done for the Bullet Cluster (see Color Plate 3). By studying the distortion in background galaxies at different redshifts, a 3D map of the matter (dominated by cold dark matter) could be made. Their survey was very narrow (1.3° × 1.3° on the sky) and very deep (out to a redshift of 0.9). Their paper (Massey et al. 2007) was titled "Dark matter maps reveal cosmic scaffolding"; they found the matter arrayed on "a loose network of filaments, growing over time, which intersect in massive structures at the locations of clusters of galaxies." The ESA/Hubble press release of January 7, 2007 said: "The map is consistent with conventional theories of how structure formed in the evolving Universe under the relentless pull of gravity, making the transition from a smooth distribution of matter into a sponge-like structure of long filaments."

Meanwhile we were using a new tracer to follow structure formation in the Sloan Survey, LRG galaxies. The LRGs are large, very luminous, elliptical galaxies that are characteristically red in color because of their lack of new star formation. Because the LRG galaxies are so bright, they can be surveyed out to larger distances than regular galaxies, which enables us to cover larger survey volumes. The LRG galaxies, being so large, form in large cold dark matter halos, making it particularly easy to model their formation in an *N*-body simulation that follows cold dark matter and uses a biased galaxy formation algorithm. Figure 9.9 shows a genus curve of the luminous red galaxies (LRGs) from the Sloan Survey, which we (Yun-Young Choi, Park, Kim, and I) studied in 2009. The wide-angle survey includes galaxies out to a redshift of 0.44; that is, the observed wavelengths ($\lambda_{observed}$) of their spectral lines have been shifted to the red, to wavelengths longer than their wavelengths seen in the lab on Earth by a fractional amount $(\lambda_{observed} - \lambda_{lab})/\lambda_{lab} = 0.44$. This shift is due to the Hubble expansion of the universe. The more distant the galaxy, the higher the redshift. We thus include galaxies up to 4.8 billion light-years from us. The topology is studied with a smoothing length of 165 million light-years.

The jagged genus curve shows the observation results for the LRGs: about 80 donut holes are seen in its spongelike median density contour (the peak of the curve at the center). The curve follows the general W shape of the Gaussian random-phase curve but shows more isolated clusters (about 40) than isolated voids (about 30). This excess of clusters over voids is expected due to nonlinear gravitational effects as first explained by Matsubara (described earlier in the discussion of the Sloan Great Wall). Biased galaxy formation (the fact that galaxies are more likely to form in high-density regions of cold dark matter) can enhance this effect. The fainter gray curve gives the average of 12 mock SDSS surveys from the Kim and Park large Horizon Run *N*-body simulations; it shows what we should expect to find, on average, from our simulations. Comparing the gray curve with the jagged curve, we found that the simulations show the same excess of isolated clusters over voids as found in the observations. No free-fitting parameters are applied to the simulations at all. Even the amplitude of the genus curve is predicted with remarkable accuracy. The agreement between the simulations and the observations—to within the observational errors—is

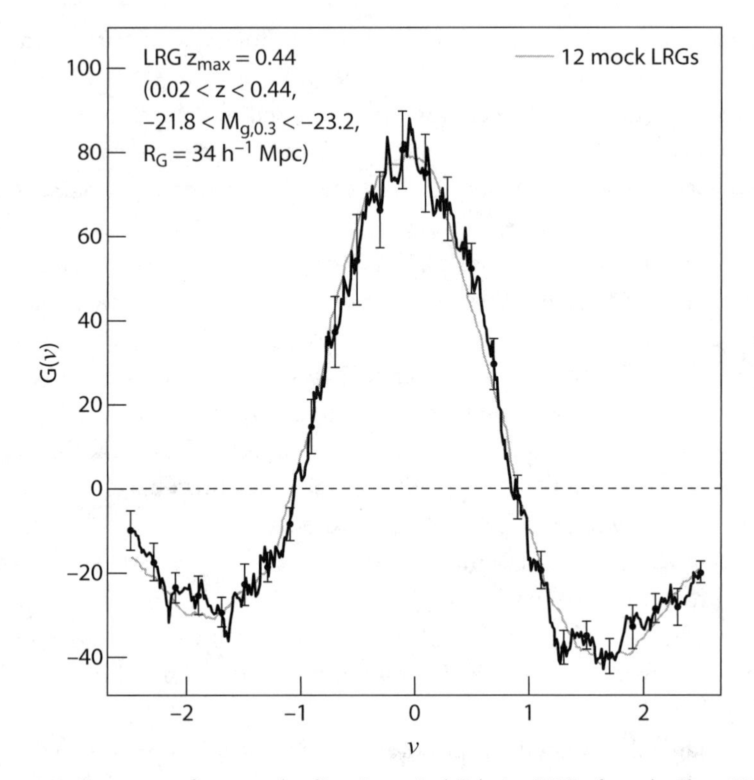

Figure 9.9. Genus curve for a sample of Luminous Red Galaxies (LRGs; from the Sloan Digital Sky Survey) out to a redshift of $z = 0.44$ (jagged line) is compared with the mean of 12 mock surveys from a large Cold Dark Matter computer simulation (faint gray line). Both are smoothed with a smoothing length of 165 million light-years. (Credit: Adapted from: J. Richard Gott, Yun-Young Choi, Changbom Park, and Juhan Kim, *Astrophysical Journal*, 695: L45, 2009)

remarkable indeed. Recall that the simulations are started with Gaussian random-phase initial conditions; their success, therefore, supports the proposition that the initial conditions were Gaussian random-phase as predicted by inflation.

In the last several years, the Sloan LRG survey has been extended to larger redshift in a continuation of the survey known as Sloan III. This allows us to study the topology at still-larger scales. The latest genus curve produced by my student Prachi Parihar, Michael Vogeley, Yun-Young Choi, myself, and the members of our topology group is shown in Color Plate 12, along with 12 mock surveys from the Horizon Run 3 simulation by Kim and Park.

Can you tell which one of the 13 jagged curves shows the observations? Give up? It's the dark blue one. We used a smoothing length of 102 million light-years. It represents the largest volume surveyed to date and contains approximately 500 donut holes in its multiply connected, spongelike, median density contour. The fact that these simulations agree so well with the observations in both amplitude and shape of the genus curve is a great victory for the standard model of inflation. The largest structures in the universe are giant filaments of galaxies stretching between the great clusters—the greatly expanded fossil remnants of initial random quantum fluctuations. Grown by gravity with the help of the mysterious cold dark matter, they form a magnificent, spongelike cosmic web.

Chapter 10

Spots in the Cosmic Microwave Background

So far, our study of 3D topology of galaxy clustering has extrapolated backward to deduce something about the random-phase nature of the initial conditions. But what if we could see the initial conditions directly? The cosmic microwave background, which comes to us from an epoch just 380,000 years after the Big Bang, allows us to do just this. When we observe the cosmic microwave background, we are seeing out to a spherical shell (where we hit the fog of the early universe), which has a radius of about 13.8 billion light-years. We are seeing directly the initial conditions of the universe, just 380,000 years after the Big Bang. It's like looking directly at Einstein's baby picture, as opposed to doing a computer age regression of a picture of him as an adult, to figure out what he might have looked like as a newborn baby. Because of the finite velocity of light, when we look out in space to the fog bank about 13.8 billion light-years away, we are also looking back in time, to the epoch of recombination about 13.8 billion years ago, just 380,000 years after the Big Bang. At that time, on the largest scales density fluctuations were about 1 part in 100,000, creating temperature fluctuations of about 1 part in 100,000 in the cosmic microwave background as one looks from place to place in the sky. These density fluctuations grow into clusters of galaxies by the action of gravity over the course of the next 13.8 billion years. Standard inflation predicts Gaussian random-phase initial density fluctuations and, therefore, Gaussian random-phase temperature fluctuations for the cosmic microwave background. These are waves, with amplitudes picked from a Gaussian distribution (a bell-shaped curve), going in all possible random directions in the sky with random phases (random locations of their peaks and troughs). Just as a Gaussian random-phase distribution in three dimensions implies a 3D spongelike topology, a Gaussian random-phase distribution in two dimensions (on a map of the sky) implies a particular 2D topology.

Color Plate 13 is a temperature map over the whole sky of the cosmic microwave background made by Wes Colley and me, along with our colleagues, using data from the WMAP satellite. Looking out in all directions, the sky looks like a sphere. The night sky looks like a hemispherical dome overhead, but Earth blocks out the other half of the sky. If Earth were to disappear suddenly from under your feet, you would find yourself floating in space, and if you looked down you would see another hemisphere of stars below you. This completes the celestial sphere. The WMAP satellite scans the entire sky, covering all the cosmic microwave background. This map shows the inside surface of the celestial sphere. In order to plot this on a piece of paper, one has to project the spherical sky onto a flat map. One of the standard map projections used for mapping the spherical Earth onto a flat map can be adapted to plot the celestial sphere. The WMAP team chose the Mollweide equal-area projection to do this. Perhaps you have seen such a map of Earth. A Mollweide projection of Earth shows its entire surface as an elliptical map, with the North Pole at the top of the ellipse, the South Pole at the bottom, and Earth's equator as a horizontal line stretching across the middle of the map. In the same way, a Mollweide projection of the celestial sphere plots the entire sky as an elliptical map, with the North Galactic Pole at the top, the South Galactic Pole at the bottom, and the galactic equator as a horizontal line spanning the center of the map. If we show the visible sky on such a projection, you will see stars sprinkled over the entire map, with the faint band of the Milky Way as a horizontal band crossing the center of the map from left to right. But the cosmic microwave background map is a radio map, a map showing radiation only from the radio portion of the spectrum; furthermore, radio radiation from the Milky Way has already been carefully subtracted out. Radio radiation from the galaxy has a different spectrum from the thermal spectrum of the microwave background; the WMAP team observed at different frequencies so that the galactic radio radiation could be estimated at each location and subtracted out. Thus the WMAP team produced a map of the cosmic microwave background alone. The Doppler effect due to the motion of Earth relative to the cosmic microwave background has also been subtracted out. Temperature fluctuations in the cosmic microwave background of order 1 part in 100,000 can then be seen.

The temperature fluctuations we see in the cosmic microwave background arise from several effects:

1. A region that is hotter than average at recombination will appear hotter to us. This is straightforward. Regions that are cooler than average will appear cooler. This is important for regions smaller than 1° in angular scale.
2. A region that has higher-than-average density at recombination will create a gravitational well; photons must climb out of it. These photons will lose energy as they climb out, and get an extra redshift, making that region look cooler. Underdense regions, by comparison, will look hotter. This is called the *Sachs–Wolfe effect*, and it is important for regions larger than 1° in angular scale.
3. Additional Doppler redshifts and blueshifts occur because of the peculiar motions of matter as it is drawn into denser regions and as it oscillates due to sound waves in the early universe (baryon acoustic oscillations). These Doppler effects are important on angular scales smaller than 1°.
4. Some extra redshifting and blueshifting occurs as photons climb in and out of growing gravitational wells as they pass through the growing superclusters on their path to reach us. This is called the *integrated Sachs-Wolfe effect* because it integrates what happens to the photons on their way to us. This is important on scales larger than 1°.
5. When a cluster of galaxies lies in the line of sight, microwave background photons can be scattered by electrons in the hot gas in the cluster and kicked up to higher frequency (the *Sunyaev-Zeldovich effect*). This depletes the number of microwave background photons seen at low frequency in the direction of the cluster, creating a tiny (approximately 0.03° wide) cold spot in the microwave background. Foreground clusters can be detected in low-frequency (less than 218 Gigahertz), very high-resolution maps as rare cold spots and in high-radio-frequency (greater than 218 Gigahertz), very high-resolution maps as rare hot spots. This effect due to hot cluster gas has been ob-

served and has been used to detect massive clusters but is not important on 1° scales. If one observes at a radio frequency of 218 Gigahertz, this thermal scattering should not create a hot or a cold spot, but galaxy clusters can still produce hot or cold spots by this scattering effect if they have peculiar velocities toward or away from us along the line of sight. This is called the *kinematic Sunyaev-Zeldovich effect* and is about twenty times smaller in amplitude than the regular Sunyaev-Zeldovich effect. Nick Hand and colleagues (2012) have detected this effect statistically by comparing nearby pairs of galaxy clusters, which have—on average—peculiar velocities along our line of sight of opposite signs as gravity pulls them together—also not important on 1° scales.

6. Gravitational deflection of light along the line of sight (weak gravitational lensing) can move hot and cold spots somewhat on the sky and slightly distort them in shape, but it does not change the topology of the spots when seen at the WMAP satellite angular resolution. This lensing effect creates small but characteristic shape squeezing of the spots, which has been detected statistically.
7. Finally, there are additional redshifting and blueshifting effects from the squeezing and stretching of spacetime due to primordial gravity waves inherited from the inflationary epoch. This is a smaller effect than the others, not detected yet, and would play a role only on angular scales larger than 1.5°.

All these effects can be calculated theoretically. Effects 1, 2, and 3 give us a direct look at the initial conditions at recombination—accounting for most of the temperature structure we see on small angular scales. We will ultimately see all these effects compared with the observations when the power in the fluctuations at different angular scales is analyzed later in this chapter.

WMAP's predecessor satellite, the Cosmic Microwave Background Explorer (COBE), had been the first to observe these fluctuations. The satellite mission earned Nobel prizes for George Smoot and John Mather. But the WMAP satellite offers a higher-resolution view. Many people from the original COBE team rejoined to build the WMAP satellite.

The WMAP team presented their map with a rainbow color scheme, with red being the hottest, followed by orange, yellow, green, and, finally, blue, as the coldest. Their picture, which displayed the sky as both intricate and beautiful, has become world famous, appearing in many books and articles. (This WMAP map is shown as an inset at the top of Color Plate 8.) But it was relatively hard to find the average temperature contour (it was somewhere in the middle of the green band), and it was hard to compare hot and cold spots. What shade of yellow on the hot side, for example, corresponded to what shade of blue on the cold side? For our purposes of testing the topology of the microwave background, Wes and I devised a different color scheme. We colored the map in red, white, and blue. The average temperature contour is white. Regions that are hotter than average are colored in varying shades of red: those just slightly above the average temperature are colored light pink, with progressively hotter regions appearing in more and more saturated shades of red. The hottest regions are colored pure red. Regions colder than average are colored in varying shades of blue. Slightly below average temperature is light blue, with colder temperatures depicted in deeper blues. One can then directly compare hot and cold spots of equal magnitude. The scale is linear in ink. If a region is twice as far from the mean in either the hot or cold direction, it will, accordingly, have twice as much red or blue ink. Regions at the mean temperature (which is also the median temperature) are pure white. This color scheme for the WMAP data appears in Color Plate 13.

Measuring 2D Topology

To study the topology of these temperature fluctuations on the 2D surface of the microwave background sphere, we would first have to learn how to measure 2D topology. One day Adrian Melott called me to say excitedly that he had figured out how we could measure topology in two-dimensional slices in the universe. For the cosmic microwave background, one starts by drawing temperature contour lines in the microwave background map. Such contour lines on a 2D geologist's map usually represent lines of constant altitude. If there are mountains of high altitude, then contour lines will surround them. If there are depressions of low altitude, then contour lines will surround them also. Lakes have shorelines, for example, that are contour lines of constant altitude, and the ground beneath the lake represents a depression that the shoreline encloses. But we are measuring temperature, not altitude. High-temperature regions will be surrounded by high-temperature contour lines that encircle them; likewise, low-temperature regions will be surrounded by low-temperature contours that encircle them. We often see such temperature contours on 2D weather maps: *isotherms* (curving contour lines) where the temperature is 70°, 80°, 90°, and so on. We would be making such a "weather map" of the cosmic microwave background.

Adrian had figured out how to use temperature contours to measure the topology of these hot and cold regions in the 2D map of the cosmic microwave background sky. In 2D, we needed a new definition of the genus. Temperature contour lines in a 2D map have a curvature. If you were to drive a truck around a circular temperature contour line on a map, you would have to turn your truck by 360° (or 2π) as you went around and returned to where you started.

Now, imagine that you have a general contour curve that encloses a high-temperature (deep red) region. Imagine this occurs on a map with pixels, like the city-block grid in a city. Drive your truck around a closed contour—say, one that encloses the Empire State Building. As you circle the contour, keep the Empire State Building on your left. Circle the block, and you must make 4 left turns to get back to where you began. A generalized contour could have a complicated shape but must follow the grid lines of the rectangular street plan. (Ignore any diagonal roads like Broadway for this exercise.) You will make many left and right turns as you work your way around a complicated contour that encloses the Empire State Building, but you will find as you complete the circuit that the total number of left turns (of 90° each) will always outnumber the total number of right turns by 4. Try it on the 7%-high-temperature contour enclosing the highest temperature 7% of the pixels shown in Figure 10.1. Put the high-temperature side on your left and drive around on the path indicated by the arrows. You will make 6 left turns and 2 right turns, and $6 - 2 = 4$. This would be true of any closed high-temperature contour: (left turns − right turns = 4). You must complete a net excess of 4 left turns of 90° to make a full counterclockwise rotation of 360° as you circle the isolated hot region. By making the contour circuit follow a grid, you are simply ensuring that all the turning occurs at discrete points.

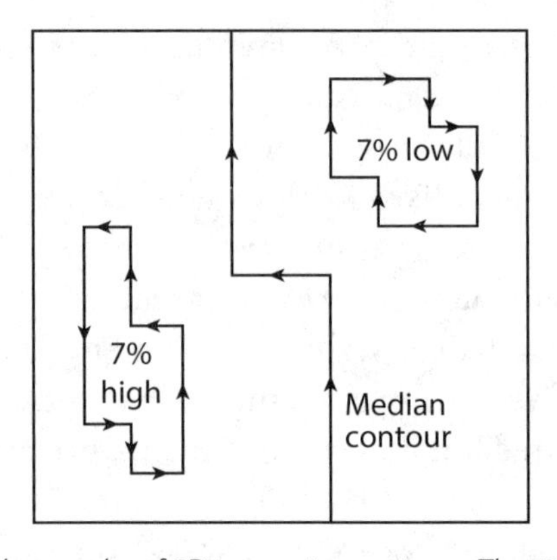

Figure 10.1. A map with examples of 2D temperature contours. The 7%-high-temperature contour encloses the 7% of the area with the highest temperature. The 7%-low-temperature contour encloses the 7% of the area with the lowest temperature. The median temperature contour divides the map in half. Drive around the 7%-high-temperature contour with higher temperature on your left; you circle counterclockwise, making 4 more left turns than right turns. Drive around the 7%-low-temperature contour with higher temperature on your left; you circle clockwise, making 4 more right turns than left turns. The median contour has an equal number of left and right turns. Counting the number of left turns minus the number of right turns required to navigate a contour enables us to calculate its genus. (Credit: J. Richard Gott)

Figure 10.1 also shows the 7%-low-temperature contour, which encloses the coldest 7% of the pixels in the plane. Finally, there is the median density contour, which divides the plane into two equal-area pieces, with the higher-temperature pixels on one side and the lower-temperature pixels on the other.

Look at all the vertices that lie on the 7%-high-temperature contour. Each of these vertices is surrounded by 4 pixels. Some of them must be hotter than the 7%-high-contour value, and some of them must be colder. If two are hotter and two are colder, you are going straight and no turn is being made. If one is hotter and three are colder, you are making a left turn at that vertex, and if one is colder and three are hotter, you are making a right turn. For this particular contour level, you can find all the vertices that fall on that contour line and count how many are surrounded by one hotter pixel and three colder pixels (left turns) and how many have one colder pixel and three hotter pixels (right turns). By obtaining the net result for left turns minus right turns and dividing it by 4, you can figure out how many isolated high-temperature regions you have. With this technique, you can count the number of isolated hot spots lying above a particular contour threshold.

Similar arguments apply to the low 7% contour. Here, if you put the high-temperature side on your left, as before, you must circle the contour *clockwise*, and your number of left turns minus right turns equals −4.

For a *general* contour, if we calculate the value (number of left turns − number of right turns)/4 required to draw the entire contour with the high-temperature side on your left, it will equal *the number of isolated hot spots minus the number of isolated cold spots.* This is Melott's definition of the genus for two-dimensional topology maps:[1]

2D genus = number of isolated hot spots − number of isolated cold spots.

The genus of the 7% high contour is +1, because it encloses 1 isolated hot spot and no isolated cold spots. The genus of the 7% low contour is −1, because it encloses 1 isolated cold spot and does not encircle any hot spots. The median density contour has 1 left turn and 1 right turn, making its genus equal to 0; it just wanders around, enclosing neither hot nor cold spots.

Hot Spots Minus Cold Spots

What do we expect the 2D genus to look like for a random slice through our sponge? (After all, the cosmic microwave background is a spherical shell that makes a 2D slice through the 3D sponge.) If the temperature fluctuations have a Gaussian random-phase distribution, as we expect from inflation, the 2D genus formula adapted from Adler (1981) is

$$\text{2D genus} = \text{number of hot spots} - \text{number of cold spots} = A \exp\left(\frac{-\nu^2}{2}\right).$$

A is the amplitude of the genus curve, and $\exp(x) = e^x$, where $e = 2.718281828\ldots$ is the base of the natural logarithms. The variable ν is a measure of the area enclosed by the temperature contour, just as it was in our 3D contours: the 7% high contour is $\nu = 1.5$, the 7% low contour is $\nu = -1.5$, and the median contour is $\nu = 0$, as before.

The genus curve is shown in Figure 10.2.

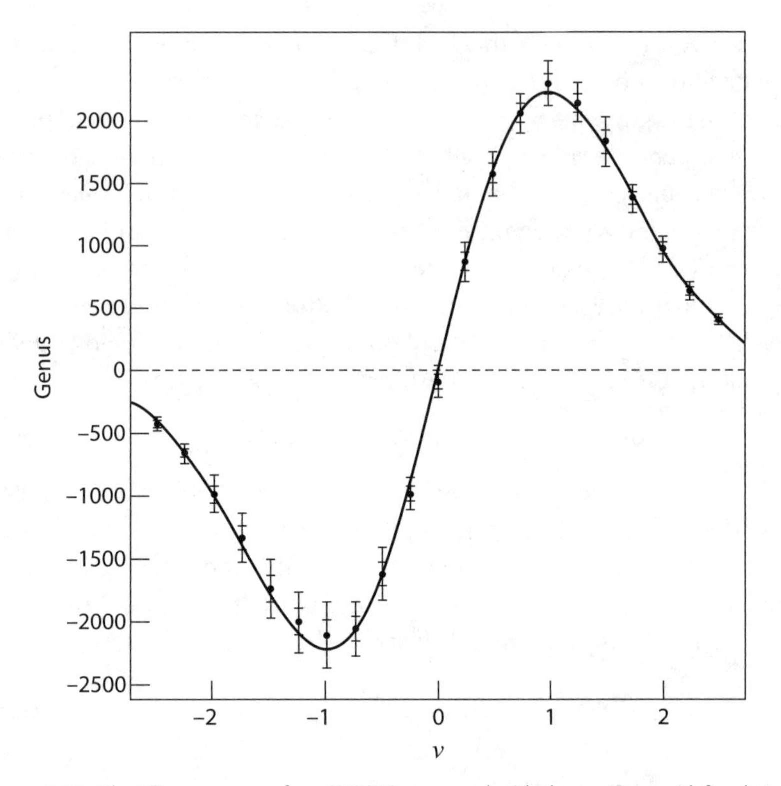

Figure 10.2. The 2D genus curve from WMAP compared with theory. Genus (defined as number of hot spots minus number of cold spots) is plotted as a function of temperature threshold. The theoretical curve for Gaussian random-phase initial conditions is a solid line. The WMAP genus data appear as dots, with 68% and 95% confidence error bars. The median temperature contour is $\nu = 0$. For temperature contours above the median, we have $\nu > 0$, and we see many hot spots; the genus is positive. For temperature contours below the median $\nu < 0$, we see many cold spots; the genus is negative. The map shows about 2,200 hot spots and 2,200 cold spots. (Credit: W. N. Colley and J. Richard Gott, *Monthly Notices of the Royal Astronomical Society*, 344: 686, 2003)

The solid curve is the theoretical formula just given and the data points with the error bars are from our analysis of the WMAP data (Colley and Gott 2003). The error bars (68% and 95% confidence limits) are calculated by looking at the observed variation in the hot-and-cold-spot counts in 12 separate regions of the sky. For a high-temperature contour ($\nu = 1.5$), which is in the red, the genus curve is positive—consistent with the large number of isolated hot spots (red spots) in Color Plate 13. For a low-temperature contour ($\nu = -1.5$), which is blue, we see many isolated blue spots, as expected. One can see at a glance that there are the same number of hot (red) spots as cold (blue) spots. Thus, the genus curve shows that the distribution of red and blue ink in Color Plate 13 is symmetric, as expected. If these fluctuations are due to random quantum fluctuations, as predicted by inflation, then red and blue (positive and negative) fluctuations must be equivalent.

The agreement between the WMAP data and the Gaussian random phase theory is extraordinary. Earlier, Changbom Park and I had helped the COBE satellite team study the 2D topology results from their lower-resolution map of the cosmic microwave background. It also fit the theoretical random-phase curve well. The higher-resolution WMAP data just show more detail and more hot and cold spots, making the error bars smaller. Departures from a Gaussian random-phase distribution can be quantified by a parameter f_{NL} invented by Komatsu, Spergel, and Wandelt (2005). A perfect Gaussian random-phase distribution would have $f_{NL} = 0$. As observations have gotten better and better, the limits on f_{NL} have gotten closer and closer to zero, honing in on the near-zero value predicted by standard inflation.[2] The cosmic microwave background looks just like what a 2D section through a Gaussian random-phase 3D cosmic sponge should look like.

1D Topology

We've covered 3D topology and 2D topology; is there any way to treat 1D topology? Barbara Ryden from our group discovered a way to do topology in a 1D rodlike survey. Here, a density contour would divide the rod into line segments that were either higher or lower than the contour. Going from left to right, one would find regions that were higher, then lower, then higher, then lower than the contour. Just as in a road trip, you would be crossing above and below a certain altitude contour. Barbara showed that the 1D topology was measured by the number of up or down contour crossings. We could define this as the 1D genus. The formula to relate this to the density enhancement of the density contour as measured by ν had already been derived in 1945 by the mathematician S. O. Rice:

$$\text{1D genus} = \text{number of up or down contour crossings} = A\exp\left(\frac{-\nu^2}{2}\right).$$

A is the amplitude and exp is the exponential function. The genus curve is a bell-shaped (Gaussian) function that is high in the center (at $v = 0$) and low on the sides as v becomes either large and positive or large and negative. That means that the median density contour has the most up or down crossings as one looks down the rod survey. If the rod is divided into equal high- and low-density parts, as one goes along the rod survey from one end to the other, one will often be either going above or crossing below the median density in a random-phase Gaussian field. Suppose one has a rodlike survey of galaxies. If one sets a high threshold, such as $v = 1.5$, which encloses the 7% high regions, one finds only a few isolated clusters and, therefore, relatively few up or down crossings are encountered. How many fewer? The answer is $\exp(-1.5^2/2)$, or about ⅓ as many up or down crossings as the median density contour.

Eventually, the 1D topology formula was checked by David Weinberg using long, rodlike samples containing neutral (nonionized) hydrogen clouds along the lines of sight to distant quasars. Each hydrogen cloud along the line of sight to a distant quasar would absorb some of the light from the distant quasar at a particular wavelength, which would be redshifted, depending on its distance from us, in accordance with Hubble's law. A series of hydrogen clouds along the line of sight to a distant quasar would absorb a series of narrow lines from the quasar's spectrum. If the hydrogen clouds were equally spaced along the line of sight, the absorption lines would be equally spaced in the spectrum. If the hydrogen clouds were clustered, then the absorption lines would also be clustered. By smoothing the spectrum, we identify higher- and lower-density regions, corresponding to dense and sparse regions of hydrogen clouds. The results for threshold crossings closely followed Ryden's Gaussian random-phase bell-shaped curve, showing that the distribution of hydrogen clouds along the lines of sight to quasars followed the 1D topology expected for a Gaussian random-phase distribution.

All these genus statistics for Gaussian random-phase distributions are related: they all start with a bell-shaped Gaussian function, $A\exp(-v^2/2)$, which is then multiplied by a polynomial that depends on the dimension of the space: $(1 - v^2)$ for the 3D genus, v for the 2D genus, and 1 for the 1D genus. These are related to the Minkowski functionals in N dimensions long known to mathematicians, as previously noted. We have found agreement with the Gaussian random-phase results using the 3D topology of galaxy clustering in the cosmic web, independently using the 2D structure of the hot and cold spots on the microwave background, and, finally, using the 1D topology of the distribution of hydrogen clouds along the lines of sight to distant quasars. All these tests support the standard picture of inflation, whereby the structures we see today originated as random quantum fluctuations in the first 10^{-35} seconds of the universe.

Power at Different Scales

Inflation also predicts the size distribution of the spots in the microwave background with remarkable precision. We can measure the *power* in the fluctuations we see (the square of the amplitude of the waves) as a function of angular scale. This is called the *power spectrum* of the fluctuations: it displays the power in the waves as a function of angular scale.

In Color Plate 14 we show the power spectrum of the fluctuations observed by the Planck satellite. The curve has many bumps in it. The main bump at a scale of about 1° is the characteristic size of the hot and cold spots we see in the picture. The other bumps in the curve are essentially overtones—like those you get when you ring a bell. These are *baryon acoustic oscillations*—due to sound waves in the early universe. The green band indicates the predictions of inflation with cold dark matter. The fit is astonishing. The difference in amplitude of the even and odd overtones in the sequence allows us to measure the amounts of normal matter and cold dark matter in the universe. The green band (prediction of the theory) broadens as we get to large angular scales, because only a few modes occur at these large angular scales, and sampling them at random from a Gaussian random-phase distribution leads to a larger random cosmic variance. The measured tilt in the primordial power spectrum relative to the Harrison–Zeldovich constant amplitude hypothesis is -0.032 ± 0.006 (68% confidence; Planck Collaboration 2015a). This compares amazingly well with the value predicted by inflation—for a simple model slowly rolling down a hill in the landscape: -0.0333.

Inflation in the early universe produces causal horizons that are only a tiny distance (10^{-38} light-seconds $= 3 \times 10^{-28}$ centimeters) away. Due to the uncertainty principle, this produces uncertainties in the geometry of spacetime: ripples, which according to Einstein's equations propagate at the speed of light—*gravity waves*. These gravity waves should produce characteristic swirls in the polarization of the cosmic microwave background radiation. So far, their detection has remained elusive. The BICEP2/Keck and Planck Collaborations (2015) has now set *upper limits* on gravity waves comparable to, but slightly below, the gravity wave amplitudes predicted from Linde's simplest version of chaotic inflation from 1983. Interestingly, a model of inflation that fits extremely well with the new Planck + Keck + BICEP2 data (cf. Planck Collaboration 2015a) is the older Starobinsky (1980) version of inflation and generalizations of it (Kallosh and Linde 2013; and Kallosh, Linde, and Roest 2013). The amplitude of gravity waves produced depends on the contour of the hill in the landscape you are rolling down. Linde assumed a valley with walls having a parabolic shape. This was based on the physics of a scalar field associated with a massive particle. Starobinsky's idea was based on one-loop quantum corrections to the vacuum state and his model has a valley with contours like an inverted bell carved out of a plateau. It produces gravity waves with an amplitude that is about a factor of 6 less than Linde's model, significantly and safely below the current observational upper limits. In the Starobinsky model, the doubling time for inflation in our universe after its formation would be a factor of 6 longer than in the Linde model; namely, the universe would be doubling in size every 3×10^{-38} seconds at the end of the inflationary epoch rather than doubling every 5×10^{-39} seconds, as in the Linde model. This sixfold-less-violent expansion would produce gravity waves 6 times smaller in amplitude. Follow-up experiments (from the South Pole Telescope, from the BICEP3/Keck experiments, and from the SPIDER high-altitude balloon experiment—all in Antarctica) as well as searches using the Planck satellite itself are in progress now. Cosmologists are watching with interest to see if these studies can open a new window on the very early universe.

Chapter 11

Dark Energy and the Fate of the Universe

In 2011, the Nobel Prize in Physics was awarded to Adam Riess, Brian Schmidt, and Saul Perlmutter for their discovery that the expansion of the universe is currently accelerating. We have seen (in Chapter 5) that the universe began with a period of very rapid expansion called inflation. We have shown that the spongelike topology of the cosmic web is consistent with the prediction of Gaussian random-phase initial conditions that inflation would provide. After the epoch of inflation ended at about 10^{-35} seconds, the universe decelerated as it became dominated by radiation and matter. The observations suggest the universe stopped decelerating about 6.9 billion years ago and has begun an epoch of accelerated expansion, which has continued ever since. Thus we have now entered another period of accelerated expansion—a form of low-grade inflation has started up again! Only now, instead of doubling in size roughly every 10^{-38} seconds, the universe appears to be doubling in size every 12.2 billion years. By following the pattern of the cosmic web as the universe expands, we can measure the properties of this late-time "inflation." Most physicists attribute this acceleration to *dark energy*, a very low-density vacuum state filling all empty space. The cosmic web is expected to play an important role in understanding the mystery of dark energy, which astronomers now believe comprises 70% of the stuff of the universe today.

The inside story of the discovery of the present accelerated expansion of the universe is told well in Robert P. Kirshner's book *The Extravagant Universe* (2002). Bob was a key participant and Adam Riess's thesis advisor. It turns out that supernovae of type Ia are good "standard candles." This means that if you observe the shape of the light curve of a supernova of type Ia, from its first appearance through its maximum apparent brightness to its dying out, you can accurately estimate its maximum *intrinsic* luminosity (in watts). There are enough supernovae of type Ia in nearby galaxies whose distances are known by other means to enable this careful calibration—from them, you can ascertain how luminous a lightbulb you are looking at in a given case. Combine this intrinsic luminosity with a supernova's observed brightness and you can determine its distance.

We now think that type Ia supernovae may be produced when two white dwarf stars, each about the mass of our own Sun, collide in a triple-star system; the shock of their collision ignites their remaining thermonuclear fuel, causing a tremendous explosion. These are some of the brightest supernovae, which are visible out to large distances. If you look out to a distance of a billion light-years, you are seeing a supernova that went off a billion years ago, when the universe was smaller than it is today. Two large teams of astronomers (one led by Perlmutter and one led by Schmidt) independently found that distant type Ia supernovae were fainter than expected, meaning that the space between them and us had been stretching faster than expected in the elapsed time since their explosion. Thus, the expansion itself was accelerating. What could be producing such an acceleration?

The most likely culprit was Einstein's original cosmological constant! Or, equivalently, it was a vacuum state of *positive* energy density and *negative* pressure, as suggested by Georges Lemaître in the 1930s—this is the view of most physicists today. (A minority of physicists suspect that the cosmological constant is, instead, on the left-hand side of the field equations as a change in the law of gravity, as Einstein originally proposed, but most physicists agree with Lemaître that it is a type of vacuum energy belonging on the right-hand side of the field equations with the other "stuff.") We call this phenomenon *dark energy*. It resembles the vacuum state found in inflation in the early universe but has a much lower density (of about 6.9×10^{-30} grams per cubic centimeter today), and it causes a low-grade version of inflation in the universe today. It is quite different from Zwicky's *dark matter,* which clusters. Dark energy has a constant energy density (6.2×10^{-9} ergs per cubic centimeter) throughout space and a pressure everywhere that is equal to the energy density but negative in sign—a sort of universal suction. Thus, the ratio of pressure to energy density, which we call w, is equal to -1.

This ensures, by the logic of Einstein's theory of relativity, that rocket ships traveling through empty space at different speeds will all measure the same amount of vacuum energy. If there were no negative pressure, these rockets would measure different amounts of vacuum energy as they flew through space. That scenario would establish a preferred velocity—the rocket velocity where the observed vacuum energy density was smaller than that seen by the other rockets. This would violate the usual idea in special relativity that in empty space there is no preferred rocket that is "at rest." Only *relative* velocities are supposed to be important in empty space. Since the pressure is uniform in space, we don't notice it. In the room where you sit, the air pressure is positive and equal to about 15 pounds per square inch. But you don't feel this pressure because it is uniform. When there are differences in pressure, such as the high- and low-pressure areas charted in a weather map, this can cause the wind to blow, and you can feel that. Pressure differences cause hydrodynamic forces. But in the case of the cosmological constant, the vacuum pressure is uniform and negative everywhere, so there are no differences in pressure, and the vacuum creates no hydrodynamic forces.

But, in Einstein's theory of general relativity, pressure gravitates as well as energy density, and negative pressure is gravitationally repulsive. Since the negative pressure operates in the three spatial directions (up-down, left-right, front-back), the repulsive effects of dark energy's negative pressure are *three* times as potent as the attraction attributable to its energy density. (Energy density is associated with the single direction of time.) The three-times-as-potent gravitational repulsion of the negative pressure therefore beats the attraction from the energy density to make the overall gravitational effect of dark energy repulsive. This repulsion is causing the acceleration of the expansion of the universe. Right now, the energy content of the universe is in three main forms: 5% ordinary matter (made of protons, neutrons, and electrons), 25% dark matter, and, remarkably, 70% dark energy. How extraordinary to discover the majority of the stuff of the universe so recently (in 1997)!

Dark energy has another curious property: as the universe expands, it retains its current value of energy density. If we had some dark energy in a box, we would have to do work to pull against the negative pressure (or suction) to expand the box to a larger size. Since the pressure is equal to the energy density, only negative, the work done to expand the box exactly matches the energy that must be added to keep the same energy density in the box as the volume of the box increases. Energy is locally conserved, as demanded by general relativity. But in the case of the cosmic expansion, this work is just being done by the adjoining "boxes" in the expanding universe—it is the work being done by the expansion itself that is keeping the energy density constant. This implies that, as the universe expands, the energy density of the vacuum will remain the same while the matter in the universe thins out. Thus, in the future, the vacuum energy density will become even more dominant over matter than it is today.

When the distances between the galaxies have doubled, the volume occupied by a given amount of ordinary matter or dark matter will have multiplied by a factor of 8, because something twice as tall, twice as wide, and twice as deep (2 × 2 × 2) occupies 8 times the volume. The density of ordinary and dark matter will have dropped by a factor of 8, because the constant amount of matter will be diluted by an eightfold increase in volume. Dark energy, by contrast, will retain its current energy density and not be diluted. Extrapolating even further into the future, the dark energy will become ever more dominant as the matter continues to thin out. The universe will continue to expand forever, as the repulsive effects of dark energy continue to accelerate the expansion. In the far future, the geometry of spacetime will approximate de Sitter space.

Theorists were quite ready to accept this remarkable picture of an accelerating universe. In fact, Jerry Ostriker and Paul Steinhardt had actually proposed including the cosmological constant to answer a couple of cosmic puzzles in 1995, 2 years *before* the discovery of the accelerated expansion of the universe. They had argued that if inflation proceeded for many doublings in the early universe, it would be expected in most cases to make the universe truly gigantic today. If that were true, its size today would be much larger than 13.8 billion light-years, making it look relatively flat on scales of 13.8 billion light-years. Inflation, by its inherently excessive stretching, should automatically make the universe very flat, by stretching out inhomogeneities. A flat universe implied $\Omega_0 = 1$ (i.e., the observed density in the universe would be essentially equal to the critical density). Guth had always emphasized this point. But if the universe consisted entirely of matter (ordinary matter plus dark matter), this would mean that the expansion was decelerating in such a way that the age of the universe would be equal to only two thirds of the observed Hubble time of 14 billion years. With $\Omega_{\text{matter}} = 1$, that would make the universe only about 9.3 billion years old, somewhat younger than the oldest stars we have found in the galaxy. It was embarrassing. Furthermore, the observations of the dark matter content of the universe stubbornly remained at about $\Omega_{\text{matter}} = 0.3$, not enough to produce a flat universe. However, if we had $\Omega_{\text{matter}} = 0.35$ and $\Omega_{\text{dark energy}} = 0.65$, then $\Omega_0 = \Omega_{\text{matter}} + \Omega_{\text{dark energy}} = 1$, they argued, making the universe flat and the age of the universe about 14 billion years, older than the oldest stars. Dark energy did not cluster with galaxies and so it would not show up in studies of the amount of dark matter. This was the Ostriker-Steinhardt theoretical argument in 1995. Krauss and Turner (1995) and Peebles (1984c) had made similar arguments, adopting $\Omega_{\text{matter}} = 0.2$ and $\Omega_{\text{dark energy}} = 0.8$.

In 2001 the Boomerang team led by Andrew Lange, a former general relativity student of mine, used a high-altitude balloon in Antarctica to study the detailed pattern of the spots in the microwave background. From the physics of sound waves in the early universe, we know the characteristic physical size of these spots, and we know the approximate distance to the cosmic microwave background, so we can calculate the spots' angular sizes on the sky, assuming the geometry of the universe is flat. The observed angular sizes of the spots agreed with a flat geometry, which implied $\Omega_0 = 1$. Combining this with the supernova data on the accelerating universe led to values of $\Omega_{\text{matter}} = 0.3$ and $\Omega_{\text{dark energy}} = 0.7$. The 70% dark energy provided both the extra energy to make the universe flat and the repulsion necessary to explain the accelerated expansion the supernova teams were observing. The WMAP satellite confirmed these approximate values to higher accuracy, as did the Planck satellite. The ratios of the heights of the peaks in the baryon acoustic oscillations (shown in Color Plate 14), as measured by the Planck satellite collaboration, allowed them to estimate the total matter density ($\Omega_{\text{matter}} = 0.308$), which includes both ordinary matter and cold dark matter, and by 2015, they could estimate from the size of the hot and cold spots in the microwave background that $\Omega_0 = 1$ to an accuracy of half a percent, implying

$\Omega_{\text{dark energy}} = 0.692$. Correlations between observed galaxy clustering and the cosmic microwave background temperature behind the clustering (called the integrated Sachs-Wolfe effect, and described in Chapter 10) also suggested a low value of $\Omega_{\text{matter}} = 0.3$, as did the observed power spectrum of the clustering itself—two results from the Sloan Digital Sky Survey. Astronomers had thus arrived at a consensus cosmological model fitting all the observations. It included approximately 70% dark energy.

Standard Dark Energy ($w = -1$)

In the standard picture, dark energy is produced by a *field* that permeates all space (like the recently discovered Higgs field, which gives particles their mass). The value of the field associated with dark energy determines the energy density of the vacuum at that location. As the field takes on different values, the vacuum will take on different energy densities. As in inflation, if you imagine the vacuum energy density as an altitude and the value of the field as a location on a map, then you can imagine a complicated landscape, with peaks and valleys. As the value of the field changes, you change location on the map and the altitude changes, corresponding to a change in the vacuum energy density. As we discussed for inflation, imagine a ball rolling on this landscape: it will naturally roll down to the bottom of a valley, where it will ultimately come to rest. If the bottom of this valley is above sea level, the vacuum there will be above zero, and that will give us a cosmological constant that is unchanging with time. This will give a pure vacuum energy where the variable w, defined as the ratio of vacuum pressure to vacuum energy density, is precisely equal to -1. In this case, the vacuum pressure is equal to -1 times the vacuum energy density, as described before. This is $w = -1$, standard dark energy.

The dark energy here is in a pure vacuum state; in order for the energy density to be seen as the same in all reference frames, by the logic of special relativity the pressure must be equal to -1 times the energy density. The field is not changing with time, so the dark energy carries *no* kinetic energy due to motion of the value of the field (which would add to both the energy and the pressure of the dark energy and alter the value of w from its canonical value of -1 in the pure vacuum state). Also, if the field were changing, this would single out a special state of motion, a special time axis pointing in the direction in spacetime in which the value of the field was falling (i.e., in the downhill direction). But if the value of the field is unchanging, there is no special downhill direction, there is no special "rest frame," and, therefore, the vacuum must look the same to all observers regardless of their state of motion: they all see *no change* in the value of the field with time. In this case, the pressure must equal -1 times the energy density, ensuring that all observers will measure the same energy density. If the field has come to rest at the bottom of a valley, it does not change, and so $w = -1$. This is true regardless of the altitude of the bottom of the valley. If the altitude is higher, the value of the energy density will be higher and the pressure will be correspondingly more negative. But the *ratio* of the pressure to the energy density in the dark energy will still be equal to $w = -1$, with the field at rest at the bottom of a valley.

How high do we expect the mountains in this general landscape to be? Of course, Einstein's famous equation $E = mc^2$ relates energy to mass and, therefore, energy density to an equivalent mass density in grams per cubic centimeter. We expect possible energy densities in the vacuum to range between plus and minus the *Planck density*. The Planck density is the *highest* density any small piece of space can have without collapsing to form a black hole. The value of the Planck density is determined by: Planck's constant h, which governs quantum mechanics; Newton's constant G, which governs gravity; and c, the speed of light—giving $2\pi c^5/hG^2 = 5 \times 10^{93}$ grams per cubic centimeter. If you ask a physicist to make a back-of-the-envelope calculation of what the energy density of the vacuum might be, he or she would probably get the Planck density. It could be either positive or negative in sign. But the vacuum density associated with the acceleration of the universe is observed to be about 6.9×10^{-30} grams per cubic centimeter—roughly a factor of 10^{123} lower than the Planck density! It is one of the greatest misfires of an order-of-magnitude calculation in the history of science. However, string theory predicts of order 10^{500} different vacuum states spaced uniformly between plus and minus the Planck density. These correspond to different locations (valleys) in the landscape and ultimately to different microscopic geometries of seven extra curled-up spatial dimensions in addition to the three macroscopic dimensions of space that we see. These different geometries in the microscopic extra dimensions correspond to different laws of physics and different vacuum energy densities.

In explaining these small extra dimensions, the analogy is usually made to a soda straw, which has one large dimension along its length and another tiny dimension around its circumference. The soda straw is a two-dimensional surface made from a sheet of paper or plastic. If its circumference is small enough, it looks to us like a one-dimensional object, a line. But each point on the line is actually a tiny circle. Put the little circles all together and you have the two-dimensional surface of a cylinder. Ed Witten has shown that all the different versions of string theory can be united with the theory of supergravity in an overall theory called *M-theory*. In M-theory, spacetime is 11-dimensional. There is the 1 dimension of time that we know, the 3 familiar dimensions of space, and 7 curled-up microscopic dimensions of space ($1 + 3 + 7 = 11$). The soda straw has 1 curled-up dimension—shaped like a circle. If we had 2 extra curled-up dimensions, they could form the surface of a tiny donut, for example, or the surface of a tiny sphere. There are about 10^{500} ways to curl up the 7 extra spatial dimensions—various complicated, microscopic, multidimensional pretzel shapes. Each corresponds to a different valley in the landscape with a different value of vacuum energy. Your ball will roll down into one of these valleys and then get stuck at the bottom of that valley.

One of the high-altitude mountain valleys could be where our universe started out (as depicted in Figure 5.3). The vacuum energy density near the end of the epoch of inflation is estimated to be roughly 10^{-10} of the Planck density. This means that the timescale for inflation to double the size of the universe is about 10^5 times the Planck time (which is 5×10^{-44} seconds—the shortest time you can measure without your clock collapsing into a mini black hole). This leads naturally to quantum fluctuations of order 10^{-5}, which we see in the cosmic microwave background. Since those early times, we have rolled down to nearly sea level. Sea level represents a vacuum with zero energy density.

In the standard picture we are currently stuck at the bottom of a valley slightly above sea level with a small positive vacuum energy density of 10^{-123} of the Planck density. The entire region of the universe we can see has a positive vacuum energy density of roughly 10^{-123} of the Planck density. Far beyond our patch of the multiverse, there may be regions where the ball has rolled down into a different valley; this will make a different pocket universe with a different value of the vacuum energy and different local laws of physics (because the values of the fields permeating the vacuum have different values there). In the infinite multiverse, *all* the different values of vacuum energy may be explored in distant regions. These are far beyond our causal horizon—the space between us and these other pocket universes is stretching so fast that the light from them can never reach us.

We could be in any one of these 10^{500} different valleys, and about 10^{377} of them will lie between 0 and 10^{-123} of the Planck density. Steven Weinberg (1987, 1989) argued that values much higher than 10^{-120} would make the universe start a rapid accelerated expansion before the epoch of equal matter and radiation density, so that fluctuations would not grow and galaxies would never form. No galaxies, no stars, no intelligent observers—we could not live in such a universe. In the multiverse, where there are many universes, some habitable and some not habitable, we are logically required to live in one of the habitable ones (Brandon Carter's *anthropic principle*). Brandon Carter noted that our observations are constrained by the fact that we are intelligent observers. Some places in the multiverse may have vacuum states that allow intelligent life, while others do not. Logically, we could not be observing from one of the regions that was uninhabitable! Thus we could find ourselves living in only some the 10^{500} different vacuum states (different valleys) from string theory.

If the bottom of the valley were significantly *below* sea level (like Death Valley), the vacuum state would cause the universe to recollapse quickly before intelligent observers had a chance to form, and we could not find ourselves living in such a universe either. A negative vacuum energy is accompanied by a positive vacuum pressure if $w = -1$ and its overall effect is gravitationally attractive. Barrow and Tipler have argued that such a negative vacuum energy universe would have to have a vacuum energy between -10^{-123} and 0 in order to give main sequence stars time to form and produce intelligent observers on circling planets before the recollapse was completed. Thus, following Weinberg's logic, habitability arguments (for developing intelligent life) require the vacuum energy in the universe today to be between -10^{-123} and $+10^{-120}$ of the Planck density. This argument was made before the acceleration of the universe was even observed! Thus, not surprisingly, we now discover that we live within this range, slightly above sea level, with a small positive vacuum energy density of roughly $+10^{-123}$ of the Planck density, or 6.9×10^{-30} grams per cubic centimeter. It is natural to suppose in this picture, where there are many (10^{380}) habitable valleys, that we may have already come to rest at the bottom of one such low-altitude valley.

Standard Dark Energy ($w = -1$) in the Future

What does this model predict will happen in the future? Some extraordinary things. Since we are at the bottom of a valley, the ball has come to rest: the field and vacuum energy are not changing. Because of the negative pressure associated with the positive vacuum energy density, the overall effect is gravitationally repulsive, and the expansion of the universe continues to accelerate. The vacuum energy stays at the same value, while the matter in the universe continues to dilute as the universe expands; the vacuum energy density comes to dominate the total energy density in the universe, soon reaching 99%, then 99.99%, and so on. The universe will double in size every 12.2 billion years. This is a rapidly expanding sequence: 1, 2, 4, 8, 16, 32, 64, 128, 512, 1024, and so forth. Every 122 billion years, the universe will expand in size by a factor of about 1,000. Just 244 billion years from now, it will be a million times bigger.

By that time, the cosmic microwave background will have redshifted to a temperature of 2.7 K/1,000,000, or 2.7×10^{-6} K, as the photons have their wavelengths stretched by a factor of a million by the expansion of space. Distant galaxies, like those in the Coma cluster, will flee from us, going faster and faster, approaching the speed of light. Their photons will become very redshifted. It will look like they are falling into a black hole. In fact, the space between us and them will be stretching so fast that the light they emit after a certain point will never be able to cross the ever-widening gap between us. They will pass beyond our causal horizon. If friends on a distant galaxy signal us, saying *WE ARE DOING FINE*, we will receive the signals *WE* and *A . . . R . . . E*, dragged out over an infinite amount of time. But we will never receive the signals for *DOING FINE*. Einstein's special theory of relativity says that a rocket can never pass my location at a speed greater than the speed of light, but it says nothing against the *space* between two galaxies stretching so fast that a light beam carrying a signal can never cross the ever-widening space between the two. We have talked about this effect before when we were talking about particles saying goodbye to us due to inflation in the early universe. Dark energy today is causing the start of another inflationary epoch. But because it is a low-grade version of inflation, the universe doubles in size only every 12.2 billion years, instead of every 10^{-38} seconds. The doubling time now is long because the vacuum density today is so low (6.9×10^{-30} grams per cubic centimeter). In the early universe, by contrast, the vacuum density was very high (roughly 9×10^{81} grams per cubic centimeter) and the doubling time was very short.

This accelerated expansion going on today creates an observer-dependent event horizon. An *event horizon* surrounds us (as observers), beyond which lie events that *we* can never see. Stephen Hawking and Gary Gibbons showed that such event horizons will cause *us* to see thermal radiation. This is similar to the Hawking radiation emitted from black holes, which causes them to evaporate eventually. In that case, however, all observers external to the black hole agree on the event horizon of the black hole—it is observer independent. The event horizon of the black hole marks the point of no return. If a spaceship ventures inside the black hole event horizon, it can never get back outside. A black hole is a hotel where you can check in but you can never check out. Outside observers can never see any events occurring inside the event horizon of the black hole. Since all observers outside the black hole agree on the event horizon of a black hole (i.e., they all agree on the events they can't see), the Hawking radiation is real and causes the evaporation of the black hole. In the case of the exponentially expanding universe, the radiation seen depends on the motion of the individual observers. We will be at rest with respect to the Gibbons and Hawking radiation we observe. This radiation will look like the cosmic microwave background, but it would be uniform at an unchanging temperature of 7×10^{-31} K. The photons from this radiation have a wavelength of roughly 22 billion light-years. That's approximately the distance from us out to the event horizon, which also stays at a constant distance from us (like the causal horizon in inflation). The value of this Gibbons and Hawking temperature is dictated by the value of the vacuum energy density, which stays constant. About 840 billion years from now, the cosmic microwave background photons become redshifted to a lower temperature than the Gibbons and Hawking radiation and become unimportant. The temperature of the universe bottoms out at 7×10^{-31} K, the Gibbons and Hawking temperature. That's just above absolute zero on the Kelvin scale, about 4×10^{32} times colder than ice water. Still, this is a finite temperature, and as we shall see this has interesting consequences at late times. Eventually, one would come into thermal equilibrium with this thermal radiation, and there would be a heat death for intelligent life.

The stars will have all burnt out by about 10^{14} years from now. Protons should decay into positrons and neutrinos, somewhere between 10^{34} and 10^{64} years from now. Then the heaviest particles left around will be electrons and positrons. They will be so spread out that they can't find each other to annihilate. They will stay around. About 10^{100} years from now, galactic-mass black holes will evaporate by Hawking radiation and blink out in a burst of glory. Then it will be dark, with nothing to look at but the Gibbons and Hawking thermal radiation from the event horizons of the universe as a whole. It would be like watching static on your television.

Now we need some even larger numbers to help us. The number 10^{100} is called a *googol*. We can also write it as 10^100. Now 10 raised to the googol power is 10^10^100 and we call that a *googolplex*. If you wait long enough, that random static you see in the Gibbons and Hawking radiation will eventually show something interesting (not that anyone would likely be around to observe it). (Remember the old story: if you have monkeys randomly typing and you wait long enough, one of them will eventually produce a copy of one of Shakespeare's plays by chance.)

Boltzmann Brains

Once in a great while—once in every 10^10^70 years—you will see something called a *Boltzmann brain*; that is, something as complicated as the human brain will appear at random from the thermal radiation. I have argued (Gott 2008) that although you may see such a brain in the distance if you wait long enough, it would not be a self-conscious intelligent observer—because the thermal radiation is observer dependent.

You see thermal radiation that is at rest with respect to *you*. If an astronaut were passing you at high velocity in a rocket at the exact time you were observing the Gibbons and Hawking radiation, he would see Gibbons and Hawking radiation that he thought was as rest with respect to *him*. You and the astronaut are seeing *different* Gibbons and Hawking photons. You and the astronaut are dredging different photons out of the quantum vacuum state. The astronaut would not see the Boltzmann brain that you observe. The Gibbons and Hawking calculation just tells you what particles a detector would see; I have worked on such calculations with Li-Xin Li. The quantum vacuum state for the Gibbons and Hawking radiation looks like a small cosmological constant, a little additional energy density and negative pressure that is real—observers can agree on that and it has a real gravitational effect. But the only photons from the radiation that are *real* are those detected by a detector consisting of *real* particles such as electrons or positrons. For example. I might see a Boltzmann brain in the distance and take a photograph of it. The photons I detected and recorded in my camera would be real, but the Boltzmann brain would not be (because of the observer dependence).

This becomes important if the universe expands *forever*; because even though Boltzmann brains would be seen infrequently, there would be an infinite number of them, in the infinite future, and the Boltzmann brains would far outnumber normal intelligent observers made of atoms circling normal stars. That would make *us* special, and that's not allowed by the Copernican principle.[1] One might ask, if Boltzmann brains outnumber normal observers by an infinite factor, then why am *I* not one of them? Such apparent contradictions lead some physicists to rule out an infinite future exponential expansion for our universe. They think it would make us special among intelligent observers. I would argue instead that one need not worry about the Boltzmann brains: you might see one 10^10^70 years from now by chance, but *because it is not real*, you could not *be* one. It's like you went to a movie theatre and watched static on the screen for 10^10^70 years and then saw Mickey Mouse up on the screen, singing and waving at you. You would ask the usher walking by if he saw Mickey up on the screen and he would say *no*. The Mickey you see up on the screen is not an intelligent observer, asking *himself* the question, Why am I up on this screen looking down on a moviegoer and an usher? Mickey is something only *you* can see. Mickey would also not pass the Turing test for an intelligent observer (making humanlike responses to an arbitrarily long series of questions). Boltzmann brains typically disappear after a few moments. Ask Mickey one additional question and he will likely disappear before he can answer. Because you are an intelligent observer, you might *see* a Boltzmann brain pop into view in the Gibbons and Hawking thermal radiation, but you could not *be* one yourself. I ended my paper with a poem, a parody of "The Purple Cow" by Gelett Burgess (1866–1951):

> I've never seen a Boltzmann brain,
> I never hope to see one,
> But I can tell you anyhow,
> I'd rather see than be one.

Recently, Kimberley Boddy, Sean Carroll, and Jason Pollack (2015) have come to similar conclusions about Boltzmann brains based on an independent analysis.

Bubbles and Baby Universes

Up to now, we have been talking about dark energy as a vacuum energy density that is constant everywhere, but if one were to wait long enough, of order 10^10^34 years, according to a calculation by Linde, one might see a bubble of lower-density vacuum form. A bubble of finite size with a lower-energy vacuum state inside would just suddenly appear, tunneling into existence (as in Figure 5.3). Having a *lower* vacuum energy inside implies a *higher* vacuum pressure inside and the bubble will start to expand, its wall moving slowly outward at first, then faster and faster approaching the speed of light.

If such a bubble formed *within* our observable universe, it would expand forever, eventually engulfing you. Moving outward at nearly the speed of light, the bubble wall would hit you without warning, and because the laws of physics inside the bubble would be completely different from those outside (it would have a different vacuum state inside), the particles from which you were built would be destroyed or changed, likely killing you.

Linde's calculations suggest this might happen in about 10^10^34 years. But it could occur much sooner: in about 10^{136} years (according to recent calculations by Nima Arkani-Hamed, Joan Elias-Miró, and her colleagues). When one of these expanding negative-vacuum-energy bubbles hit you, it would likely kill you. According to Arkani-Hamed, the value of the Higgs field inside these bubbles would be far different from what we experience in the universe today; and since the value of the Higgs field sets the masses of elementary particles, the masses of electrons, protons, and neutrons inside the bubble would be far different. As the bubble wall hit you, the masses of these particles would be different on the other side of the wall, preventing any particles from your body from entering the bubble. You would be squashed by the wall, as Arkani-Hamed describes, "like a bug hitting the windshield of a car." Photons, which are massless, could pass inside unimpeded, however. So, from the inside, it would even look like a bug hitting a windshield. You would end with a splat!

But it is quite possible that the Higgs vacuum is stabilized against formation of negative energy bubbles by higher-energy particle physics effects we have not yet observed. A mechanism suggested by Joan Elias-Miró and colleagues involves the Higgs field coupling to a massive field like that powering inflation, which is quite plausible. Indeed, Arkani-Hamed himself thinks such a stabilization is likely. In that case, you might indeed have to wait for 10^10^34 years before a negative-energy bubble hit you—still, that is long before you would ever see a Boltzmann brain. Some physicists have cited this as another reason not to worry about Boltzmann brains. You should count only what you can ever observe and not worry about Boltzmann brains beyond your event horizon. In the far future, the real particles thin out exponentially with time so that the number of them per causal volume on average is far less than one. Thus, in the far future, most regions the size of the observable universe (within the causal horizon) will be empty of real particles and real observers. In both the 10^{136}-year and 10^10^34-year cases, the negative-energy bubbles form so infrequently that they do not *percolate*—collide with each other to make a space-filling froth. The volume of space between the bubbles continues to increase forever. Therefore, most of the volume of the universe stays in the inflating state eternally, with negative-energy bubbles continuing to form—making the universe in the far future like an ever-expanding, eternally fizzing bottle of champagne.

Another possible scenario that could occur in the far future (as suggested by Aguirre, Gratton, and Johnson (2006) and also by Andreas Albrecht) involves quantum tunneling to create a microscopic black hole (with a mass of about 1 gram) connected by a wormhole to an inflating region inside. This would look like a doorknob hanging off a door: imagine our big universe is the surface of the door, and the knob is the small inflating balloon connected to the door by a narrow stem—the wormhole. Observers living in the "door" would see only the funnel-like entrance to the wormhole, which to them would look like a black hole. The knob would have a high-density vacuum state inside, which would balloon in size, creating an entire inflationary multiverse. The microscopic black hole would quickly evaporate by emitting Hawking radiation, pinching off the narrow wormhole that connected the ballooning knob to the door. Anyone there to see the formation of such a universe would not be killed, because they would likely be far outside the region where the black hole formed. From their vantage point in the door, they would simply see the black hole form and evaporate. The burgeoning baby universe would remain hidden from them inside the black hole. Thus, as suggested by Farhi, Guth, and Guven (1990), an independent, baby, inflating universe could be formed. How long might it take before *you* were lucky enough to encounter a black hole forming a baby universe in this way? About 10^10^66 years. An expanding bubble of negative energy would likely kill you long before that happened. But since the universe continues to inflate forever in this scenario, with the volume of the inflating space between the bubbles increasing forever, the creation of such baby universes should eventually occur in the future.

According to the proponents of this scenario, it suggests yet another reason not to be surprised to find yourself to be an ordinary observer born on a planet in a universe with an inflationary beginning. Intelligent creatures can, in principle, arise in the far future directly from quantum tunneling, but only less often than every 10^10^70 years within the observable universe (because they weigh more than just a brain). Thus, new baby inflationary universes are being created at a faster rate (once

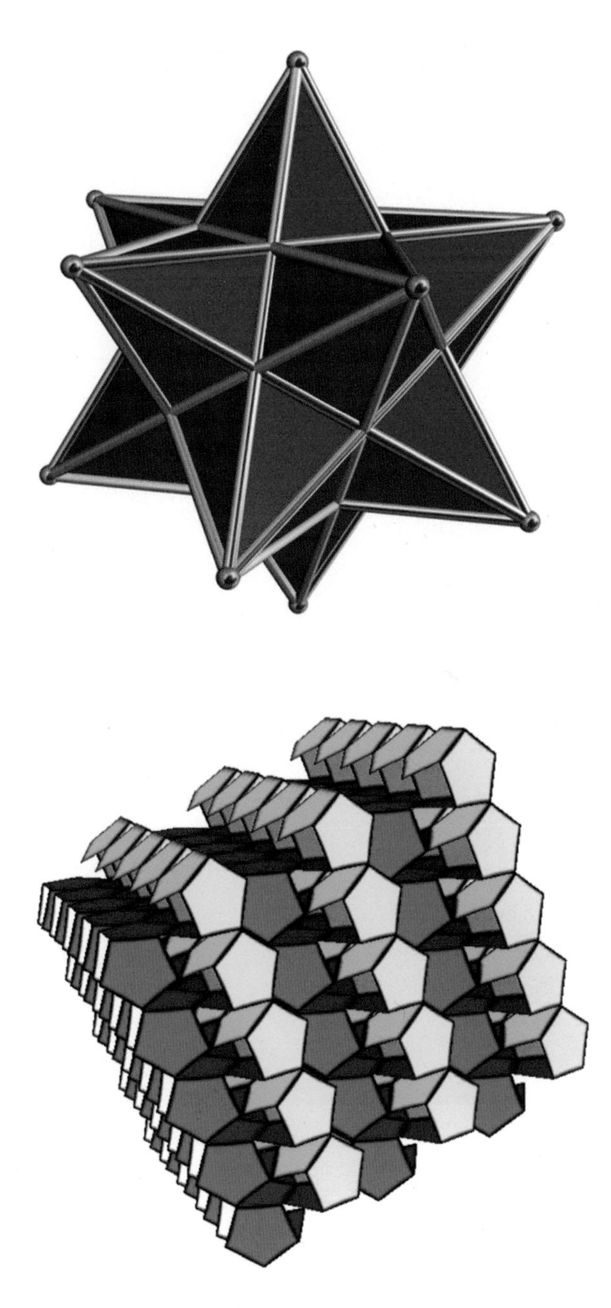

Plate 1. Kepler's small stellated dodecahedron (*top*) has 12 faces, 5-pointed stars crossing through each other. The one facing us contains 5 bright red triangles extending inward from its 5 points. My spongelike polyhedron (*bottom*) has pentagons arranged 5 around a point. Both expand the definition of a polyhedron. (Credit: (*top*) Robert Webb's *Stella* software is the creator of this image: http://www.software3d.com/Stella.php; (*bottom*) Melinda Green, http://www.superliminal.com/geometry/infinite/infinite.htm)

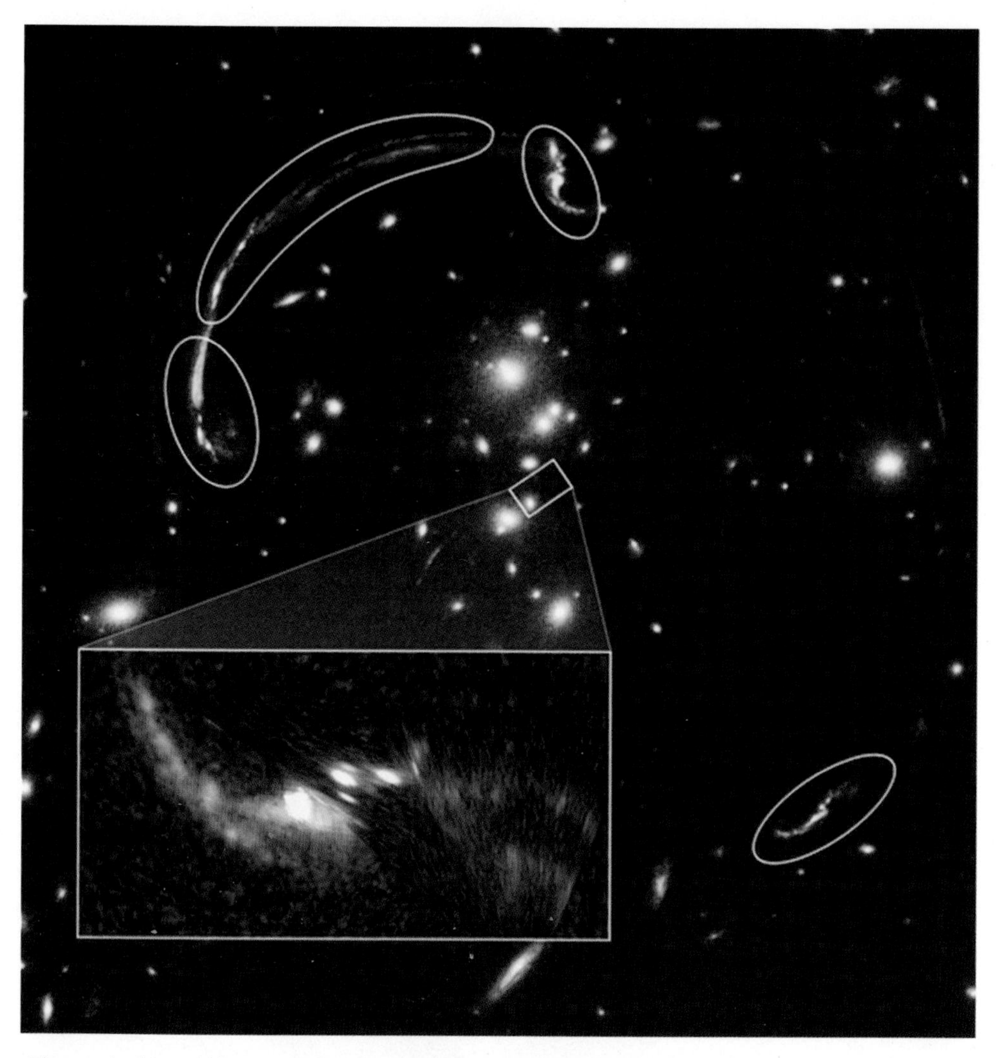

Plate 2. In this Hubble Space Telescope picture, we see a cluster of galaxies in the center (RCS2 032727-132623, about 5 billion light-years away), producing four distorted gravitationally lensed images (circled) of a background galaxy (about 10 billion light-years away) arranged approximately in a ring. The small rectangle shows the actual location of the background galaxy. The large rectangle shows its appearance from reconstructed images. (Credit: NASA, ESA, and Z. Levay (Space Telescope Science Institute))

Plate 3. The Bullet Cluster. Optical Hubble Space telescope image shows two clusters of galaxies (at left and right) that have collided and passed through each other. X-rays from hot gas appear red; dark matter, mapped by its weak gravitational lensing effects on background galaxies, appears blue. Dark matter associated with each cluster has also moved under the influence of gravity and is in two clumps just like the galaxies. Intergalactic gas from the two clusters has collided and heated forming two red clumps left behind in the center. Dark matter is not made of normal intergalactic gas--or it would be coincident with the red clumps. (Credits: X-ray--NASA/CXC/CfA/M. Markevitch et al.; Dark Matter lensing map--NASA/STScI; ESO WFI; Magellan/U. Arizona/D. Clowe et al.; Optical--NASA/STScI; Magellan/U. Arizona/D. Clowe et al.)

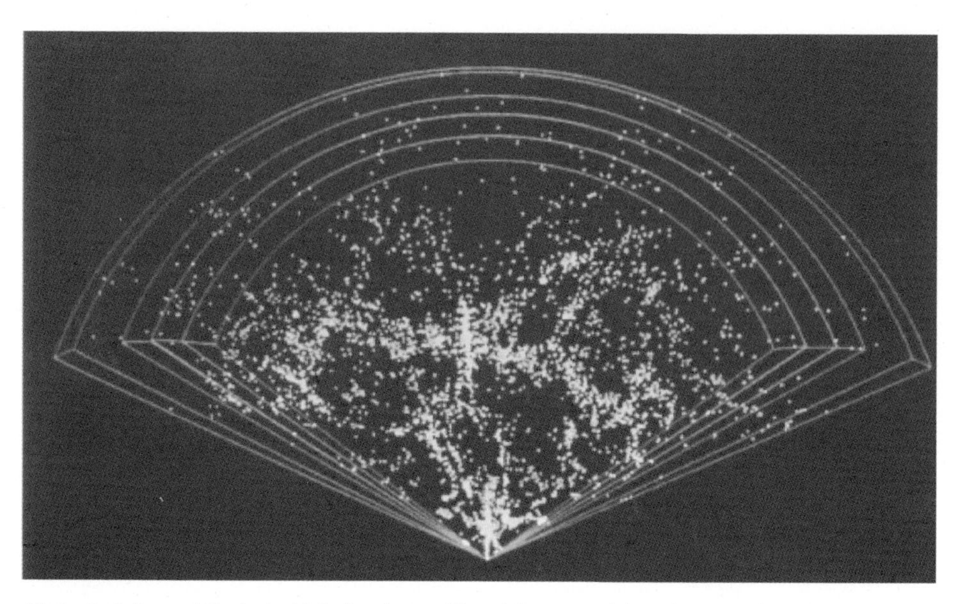

Plate 4. Geller and Huchra's thick slice (created by adding several thin slices together) shows the Great Wall of Galaxies which snakes across, from left to right. Once again, Earth is at the bottom vertex of the fans. The "dagger" at the center pointing down toward the Earth is the Coma cluster of galaxies. The various orbital velocities of galaxies in the cluster, when added to the mean recessional velocity of the cluster, smear it in the direction of the line of sight toward Earth. (Credit: M. J. Geller and J. P. Huchra, *Science*, 246: 897, 1989)

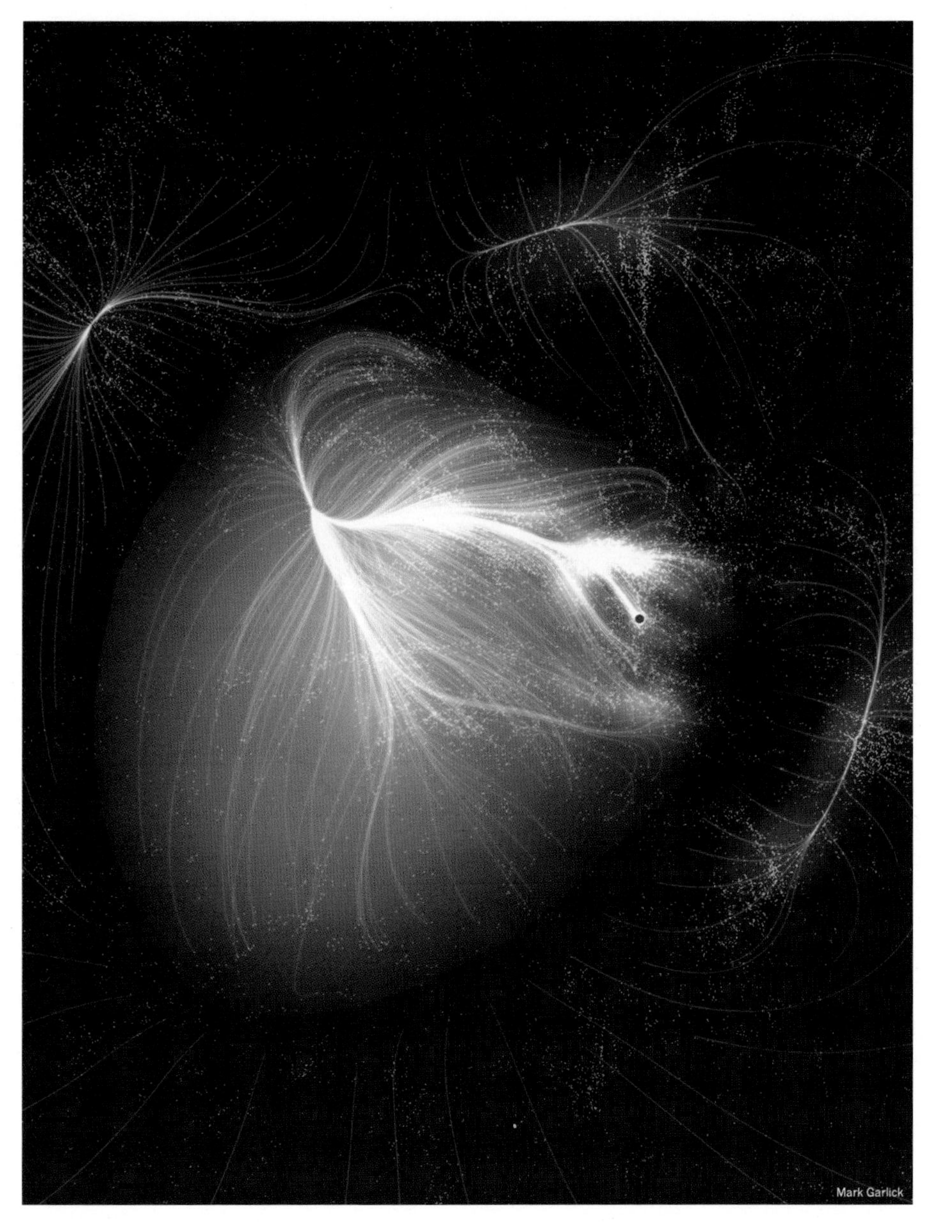

Plate 5. Velocity flow lines in the Laniakea Supercluster are shown as yellow "rivers" leading to the Great Attractor at the left. Our galaxy is the red dot at the right. The Virgo Supercluster is just above the red dot. The Laniakea Supercluster has a diameter of 510 million light-years. Flow lines in neighboring superclusters are shown in blue. (Credit: R. Brent Tully, Helene Courtois, Yehuda Hoffman, and Daniel Pomarède, *Nature*, 513 (7516): 71, 2014)

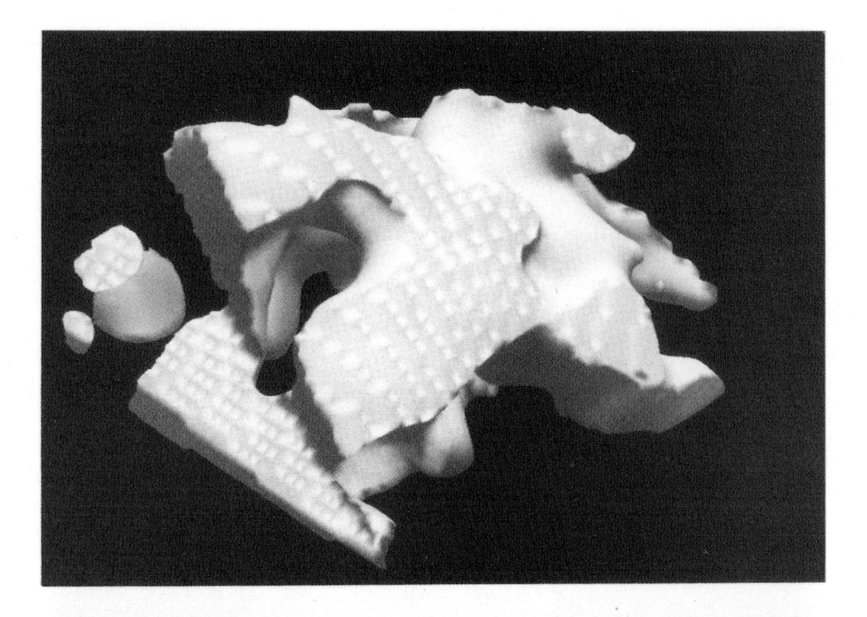

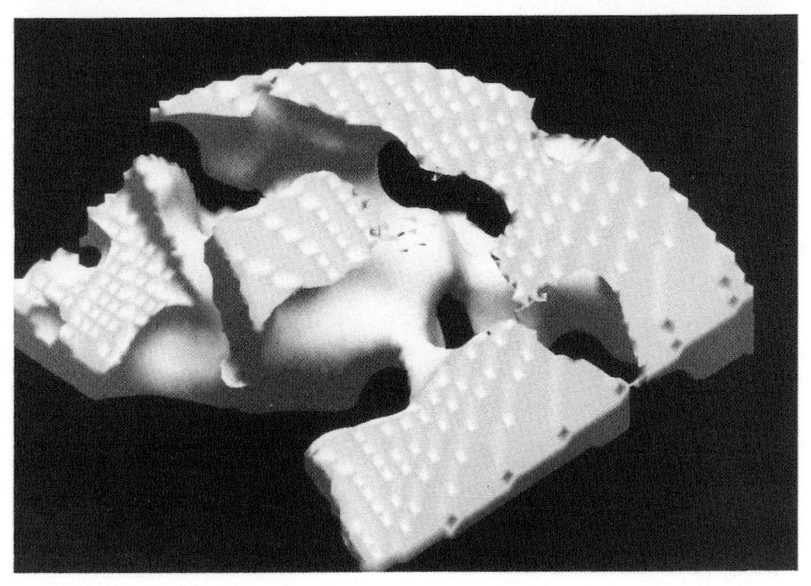

Plate 6. In this picture of the Southern CfA2 Survey, we see the 50%-high-density regions (*top*), where Michael Vogeley has placed a red light source in the middle of the large void in the middle of the survey. You can see the red light shining out into other void regions through tunnels between the voids. In the bottom panel, you can see the 50% low-density regions. If added to the high-density regions, they would produce the entire survey region. The large void in the center can now be seen as a solid blob with extensions (low-density tunnels) connecting it to other void regions. This is a spongelike topology: high- and low-density regions are multiply connected and interlocking. (Credit: Michael S. Vogeley, Changbom Park et al., *Astrophysical Journal*, 420: 525, 1994)

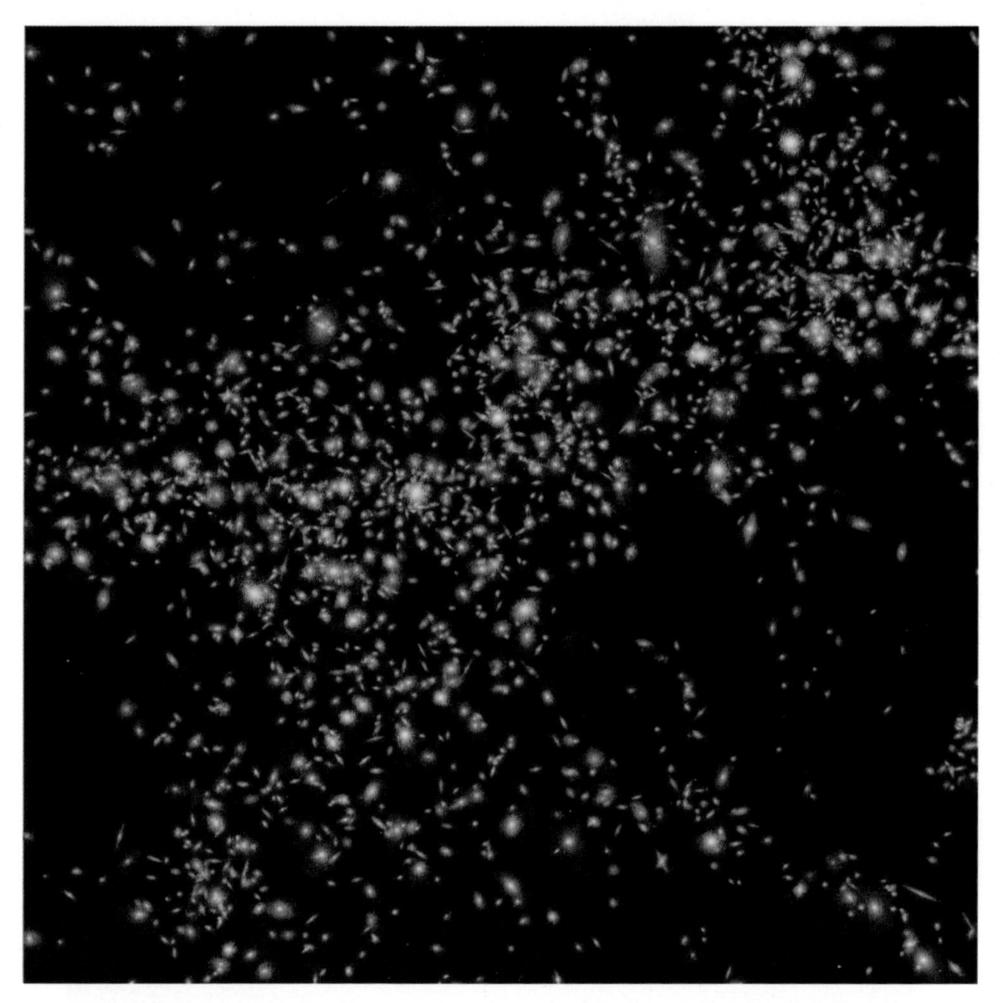

Plate 7. Eastern end of the Sloan Great Wall. Galaxies are enlarged 50x. (Credit: Courtesy of Lorne Hofstetter and J. Richard Gott)

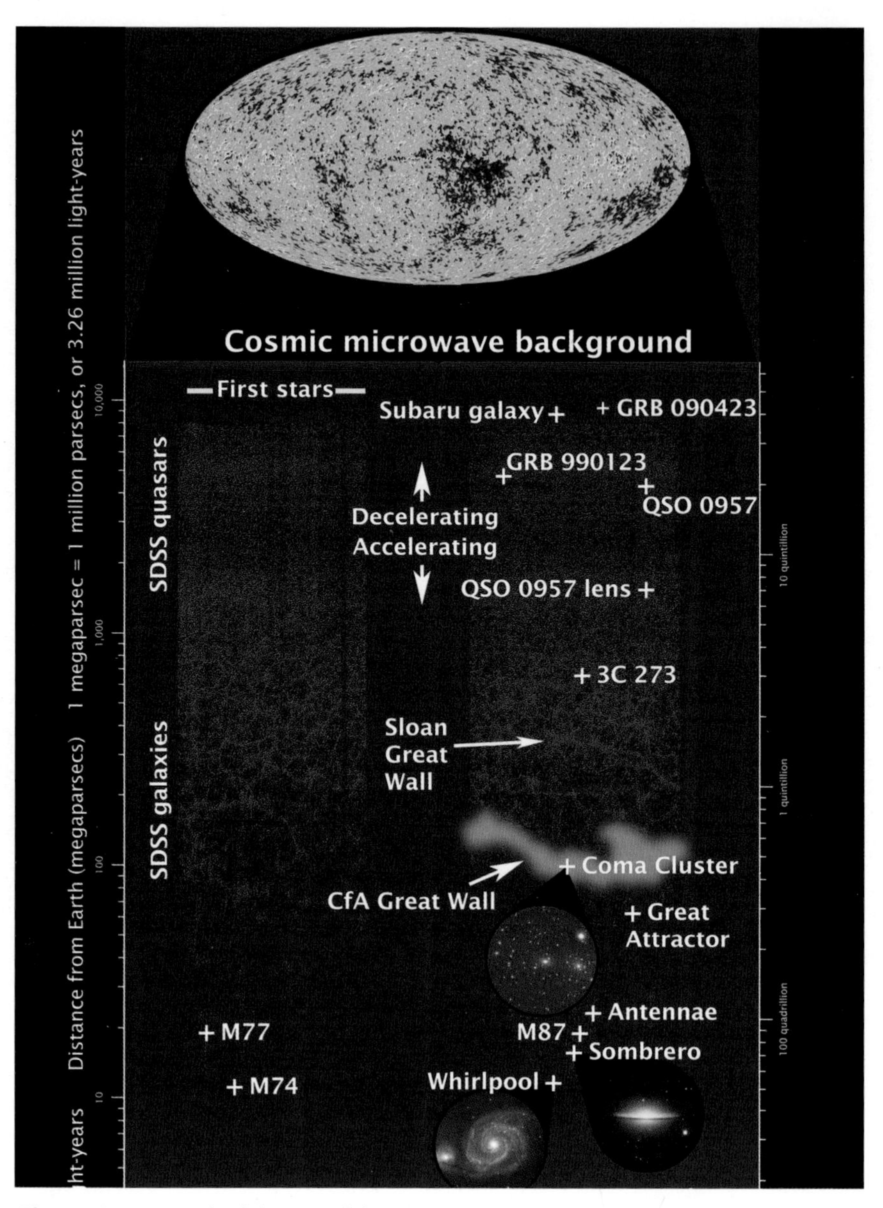

Plate 8. Top one-sixth of the Map of the Universe. From left to right, we have a 360° panorama looking out from Earth's equator, with increasing distance from Earth shown vertically. The cosmic microwave background, the most distant thing we can see, is shown near the top. The two vertical bands show the regions covered by the SDSS. Each point is a galaxy or quasar. The blank bands were not covered by the survey. Objects further away are shown at smaller scale. The Sloan Great Wall, which is three times further away than the CfA Great Wall, is shown at one-third the scale here. Actually, it is about twice as long as the CfA Great Wall. (Credit: J. Richard Gott, Mario Jurić et al., *Astrophysical Journal*, 624: 463, 2005)

Plate 9. This slice through the Millennium Run computer simulation shows the cosmic web. The scale bar in the picture has a length of 125 Mpc/h (608 million light-years), one-quarter of the Millennium Run simulation box size. (Mpc is a megaparsec = 3.26 million light-years, and $h = 0.67$ is the current value of the Hubble constant in units of 100 kilometers/second/megaparsec). (Credit: Volker Springel and the Virgo Consortium, *Nature*, 435: 629, 2005. http://www.mpa-garching.mpg.de/galform/virgo/millennium/seqD_063a.jpg)

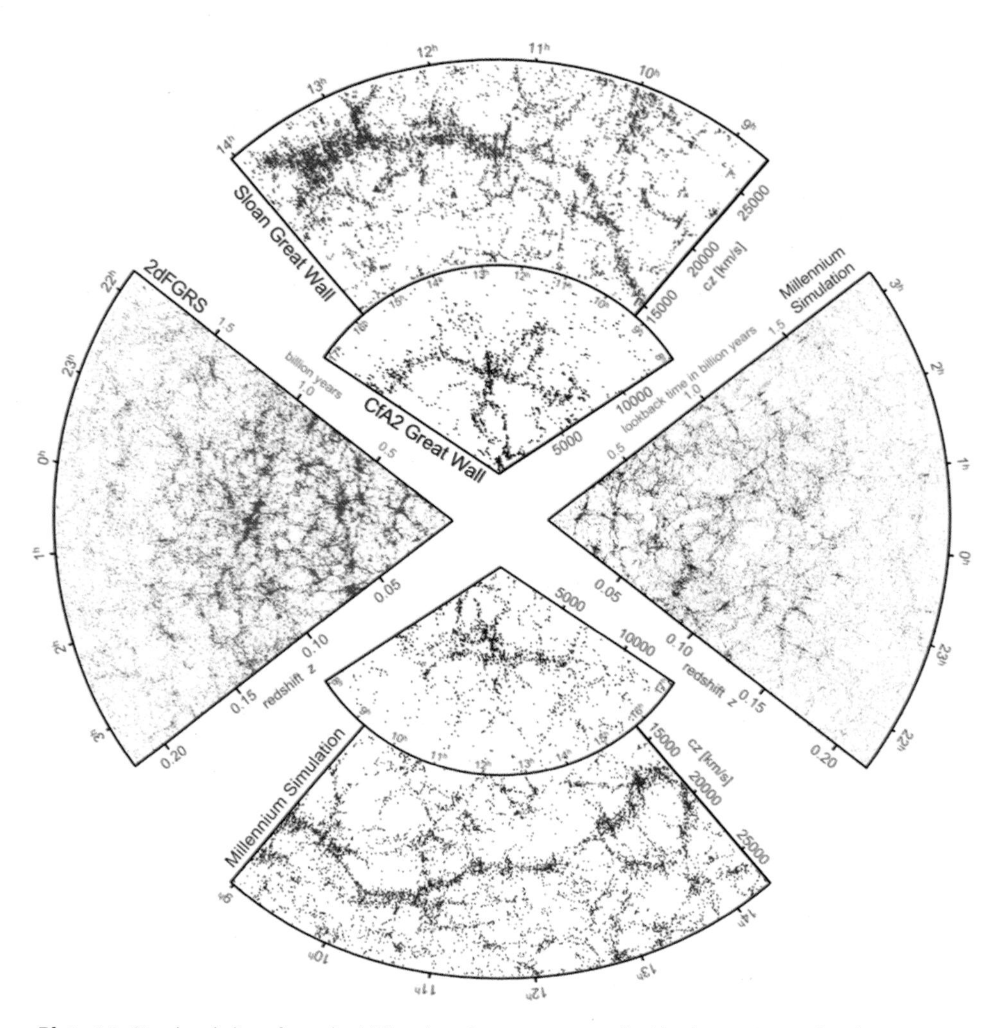

Plate 10. Simulated slices from the Millennium Run are compared with observations. The observations (in blue) are the CfA and Sloan Digital Sky surveys (*top fans*) and the 2Df survey (*left fan,* labeled *2dFGRS*). Across from these surveys are equivalent fans (in red) produced from the Millennium Run simulation, which found one great wall complex as long, but not as spectacular, as the Sloan Great Wall. (Credit: Volker Springel, Carlos Frenk, and Simon White, *Nature*, 440: 1137, 2006)

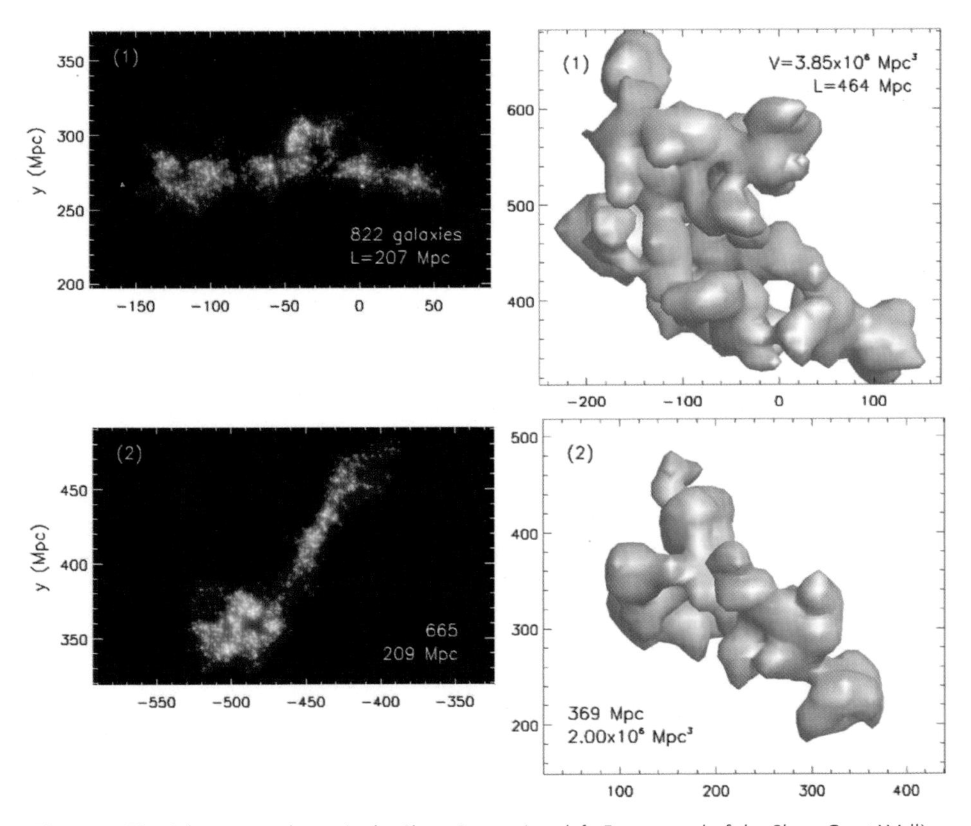

Plate 11. The richest supercluster in the Sloan Survey (*top left*: Eastern end of the Sloan Great Wall) and the largest void complex in the Sloan Survey (*top right*) are compared with the median richest supercluster (*bottom left*) and median largest void complex (*bottom right*) obtained in 200 mock Sloan Surveys made from the large Horizon Run *N*-body computer simulation. The richest supercluster and largest void complex in the observations are quite typical of those found in the simulation mock Sloan surveys. (Credit: Changbom Park et al., *Astrophysical Journal*, 759: L7, 2012.)

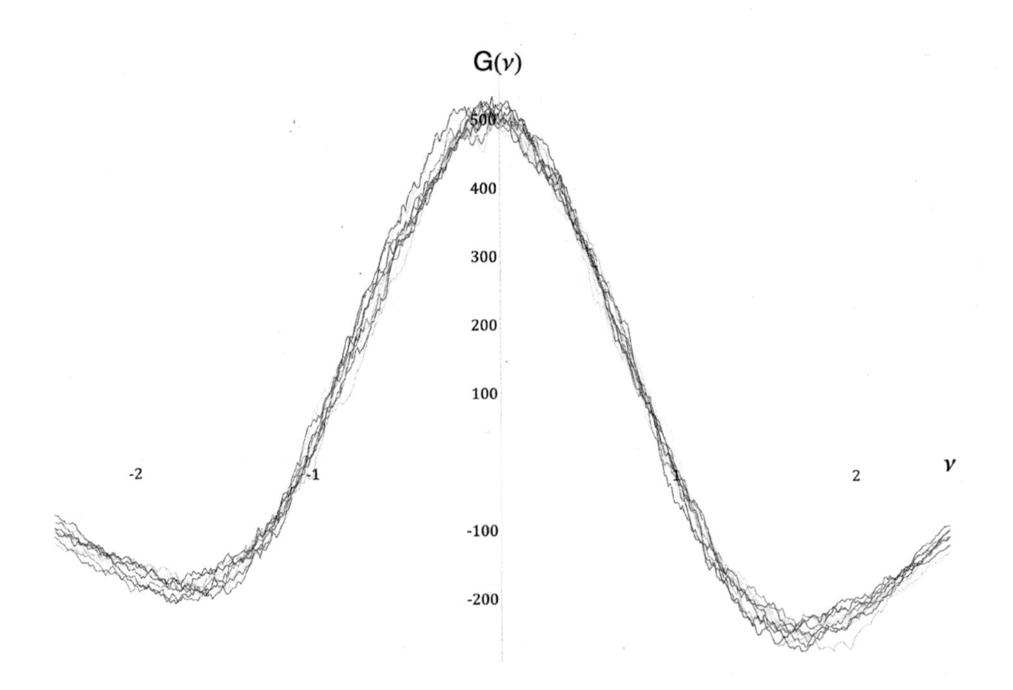

Plate 12. The genus curve (dark blue) for the observations (the BOSS survey) is indistinguishable from 12 Cold Dark Matter computer simulations (other colors) within the errors. The genus curves follow the form expected from Gaussian random-phase initial conditions, plus the slight excess of clusters over voids expected from nonlinear gravitational evolution. (Credit: Adapted from: Prachi Parihar et. al., *Astrophysical Journal*, 796: 86, 2014)

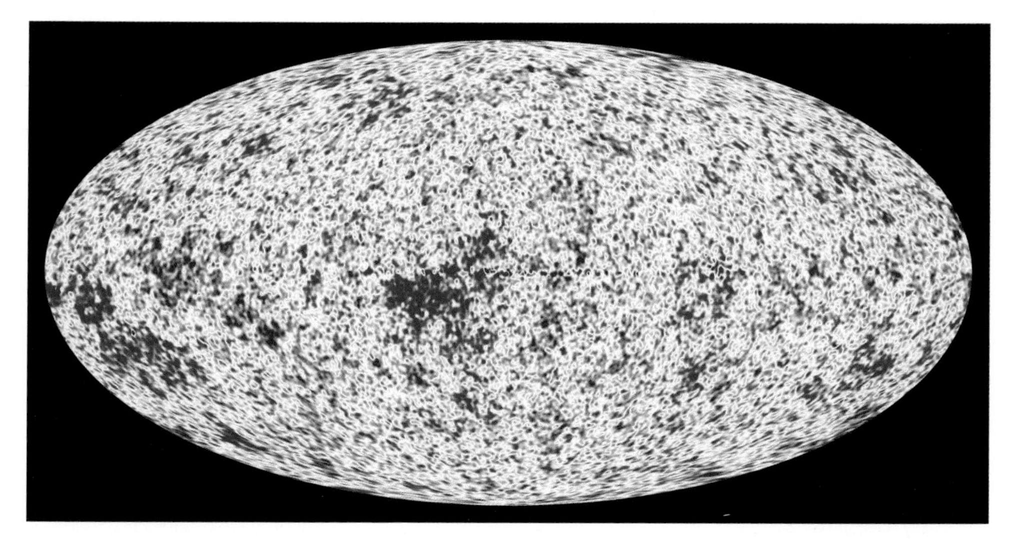

Plate 13. Temperature fluctuations obtained by WMAP are displayed in this all-sky map of the cosmic microwave background. Fluctuations in temperature are of order 1 part in 100,000. On the map, red indicates hotter than average temperature; white, average temperature; and blue, colder than average temperature. (Credit: J. Richard Gott, Wesley N. Colley, Chan-Gyung Park, Changbom Park, and Charles Mugnolo, *Monthly Notices of the Royal Astronomical Society*, 377: 1668, 2007)

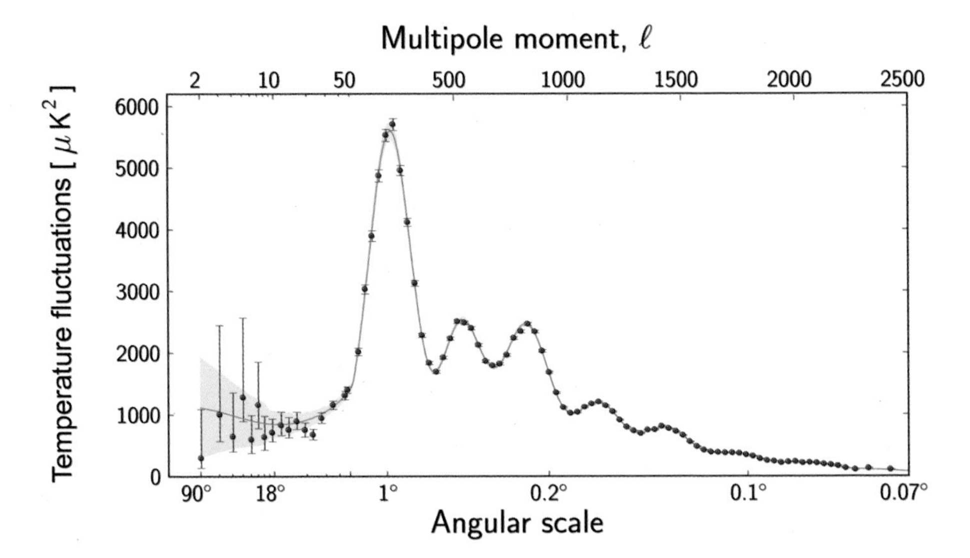

Plate 14. Power in cosmic microwave background temperature fluctuations as a function of angular scale (shown across the bottom scale in degrees). Data points (red dots) are plotted with 68% confidence error bars. Theory based on inflation appears as a green band, which becomes wider at large angular scales due to the expected statistical variation because of the small number of modes. The agreement is extraordinary. The flat portion of the curve at large angular scales corresponds to nearly constant amplitude fluctuations coming within the causal horizon due to inflation. The highest power is at a scale of 1°, corresponding to a compression sound wave in the early universe, with harmonics appearing at smaller angular scales. (Credit: ESA and the Planck Collaboration. http://sci.esa.int/planck/51555-planck-power-spectrum-of-temperature-fluctuations-in-the-cosmic-microwave-background/. Published in Planck Collaboration, *Astronomy & Astrophysics*, 571: A1, 2014)

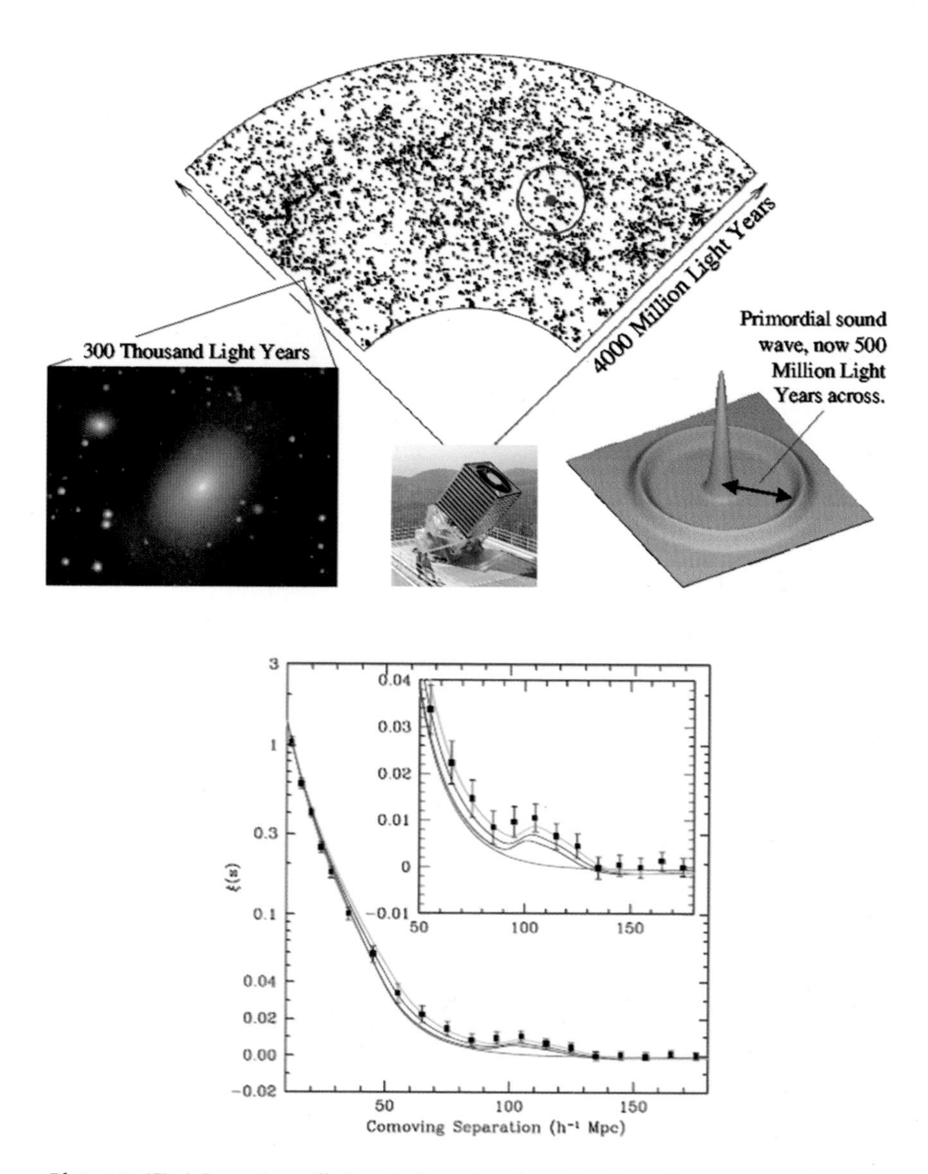

Plate 15. (*Top*) Acoustic oscillations in the early universe imprint a characteristic scale in the cosmic web. Each density enhancement today is surrounded by a bump of extra galaxies at a radius of 500 million light-years (*illustrated at right*). The SDSS telescope is shown (*center*) observing Luminous Red Galaxies (LRGs) to a distance of 4 billion light-years. A typical elliptical LRG is shown at left. Center on any LRG (red dot) and draw a radius of 500 million light-years: you will find on average more galaxies at that particular radius (the acoustic ripple). (*Bottom*) Observational data (Eisenstein et al., 2005) show the acoustic ripple as a bump in the covariance function $\xi(s)$. (Credits: (*Top*) David Schlegel, Recondres de Moriond, 20 March 2008, http://www.sdss3.org. (*Bottom*) Daniel J. Eisenstein et al., *Astrophysical Journal*, 633: 560, 2005. Courtesy of David Schlegel)

Plate 16. Our view of the cosmic web can be appreciated in this all-sky picture in the infrared from the 2MASS survey. Horizontally across the middle of the picture is our own Milky Way Galaxy in the foreground, with galaxies seen as colored dots beyond. One can clearly see the filamentary pattern of chains of galaxies connecting clusters. The most distant structures in this picture are closer than the Sloan Great Wall. (Credit: 2MASS Survey; T. Jarrett, IPAC/Caltech)

per 10^10^66 years in the visible universe) than intelligent creatures form from quantum tunneling, and each new baby, inflating universe will be producing a *large* number of new ordinary observers like you, without requiring further tunneling. Thus, new ordinary observers are being created at a higher rate than new intelligent creatures formed by quantum tunneling, so you should not be surprised to be one of the ordinary observers. Our universe might well be descended from a baby universe created in a previous low-density accelerating universe.

Besides allowing quantum tunneling, which lets you get out of your valley and, ultimately, roll further downhill, the uncertainty principle allows you, on occasion, to jump up to a higher-density vacuum state. This is like jumping up into the mountains from your position in the bottom of a valley. The bubble of high-density vacuum state that would be produced would begin collapsing right away (from its high negative pressure pulling its wall inward). But if the forming bubble were larger than the causal horizon, its interior space would be so large and inflate so fast that the collapsing bubble wall would never be able to reach the bubble's own center. In other words, the inflation inside the bubble would outstrip its collapse, and it would keep expanding. This could create a high-density inflating state that could expand forever and blow low-density bubble universes of its own. How long would we have to wait for this to happen? About 10^10^123 years. That's a really long time. But eventually there would be a recycling of universes, an idea explored over the years by Alex Vilenkin and Andre Linde.[2] Wait long enough and a high-density inflating bubble will occur that will lead to new universes like ours, where the whole process starts over. That's the standard picture for the future of our universe.

But there are other alternatives. The fate of the universe rests on the value of w, the ratio of pressure to energy density in the dark energy. If $w = -1$, we have standard dark energy as described so far. But if $w > -1$ or $w < -1$, we get two additional possibilities, *slow-roll dark energy* and *phantom energy*. These possibilities are all summarized in Table 11.1. As we shall see, studies of the cosmic web, along with more supernova studies, allow us to chart the expansion history of the universe and measure w. Let's see what the other two possibilities ($w > -1$) and ($w < -1$) imply for the future of the universe.

Table 11.1. The Value of w and the Future of the Universe†

$w > -1$	$w = -1$	$w < -1$
Slow-roll dark energy.	Standard dark energy.	Phantom energy.
Vacuum energy decreases with time.	Vacuum energy stays constant with time.	Vacuum energy increases with time.
Slower than exponential expansion.	Exponential expansion forever.	Faster than exponential expansion.
Nearly exponential expansion ends but expansion continues.	Universe doubles in size every 12.2 billion years forever.	Universe doubles in size on shorter and shorter timescales.
Temperature drops as universe becomes virtually empty.	Universe approaches a constant temperature.	Planets, then atoms, are torn apart.
	Intelligent life dies out.	
More of the cosmic web may become visible. Collisions from other bubble universes in the far future. Our universe ends in a froth of negative energy bubbles.	Bubbles and baby universes eventually form in the far future and Boltzmann brains appear even later. It's like an eternally fizzing champagne, with negative energy bubbles forming forever.	Our universe can end in a Big Rip singularity in as little as another 150 billion years.

†*Note*: $w = p_{\text{dark energy}}/\rho_{\text{dark energy}}$; it is the ratio between the pressure (p) associated with dark energy and the energy density (ρ) of dark energy.

Slow-Roll Dark Energy ($w > -1$)

Perhaps the vacuum state in the universe now is *not* sitting at rest in the bottom of a valley but rather slowly rolling down a hill. That would mean that the vacuum density in the entire universe is slowly decreasing with time. The first people to consider such a model were Peebles and Bharat Ratra in 1988—long before the accelerated expansion of the universe was even discovered. Many people have now investigated such models—which are sometimes called *quintessence* models (a name popularized by Paul Steinhardt).

The universe has undergone two inflationary epochs: (a) inflation at the beginning, in which the universe doubled in size roughly every 10^{-38} seconds; and (b) the low-grade inflation we are seeing today, in which the universe will be doubling in size every 12.2 billion years. Zachary Slepian and I, among others, have argued that the most conservative model would be one in which both epochs of exponential expansion were due to similar causes. We have observational evidence that inflation in the early universe was in a slow roll down a hill. This evidence comes from the observed slope of the power spectrum of the fluctuations seen in the cosmic microwave background by the WMAP and Planck satellites. If the accelerated expansion we see today is like that seen in the early universe, then it is also slowly rolling down a hill. We call this *slow-roll dark energy*. The field that is producing the dark energy has some kinetic energy corresponding to the kinetic energy of a ball as it rolls down the hill (proportional to the square of its velocity or rate of change with time). This kinetic energy produces extra *positive* energy density and extra *positive* pressure, which add to the positive energy density and negative pressure already in the vacuum. Thus, the total pressure in the dark energy becomes slightly less negative, while the total energy density becomes slightly more positive. This means that w, the ratio of pressure to energy density in the dark energy, has a value $w > -1$. For example, we might find a current value of $w_0 = -0.95$. The subscript 0 refers to the fact that w is evaluated at the current epoch. This is slightly less gravitationally repulsive than a cosmological constant with $w_0 = -1$ would be. The acceleration of the universe is thus slightly less dramatic than would be the case if we were sitting at the bottom of a valley. The value of w affects the dynamics of the universe's expansion via Einstein's equations. By studying the dynamical history of the universe in detail, we can measure w and determine whether the vacuum energy density is sitting at the bottom of a valley or slowly rolling down a hill. Measuring w was found to be the most important problem for astronomy to tackle in the next decade, according to the 2010 Astronomy and Astrophysics Decadal Survey conducted by the National Academy of Sciences. In analyzing the observational data, it is useful to know how w is likely to be changing with time.

If we are slowly rolling down a hill, the vacuum energy density in empty space is slowly declining with time. There is less dark energy as time goes on and the acceleration of the universe is slowing with time. Zack Slepian and I found that this follows a very simple functional form, which can be checked against the observations. The equations are like that of a ball rolling down a hill with a frictional force that is proportional to the expansion rate of the universe, because the expansion of the universe saps the kinetic energy of the field—slowing its downhill roll—just as a frictional force does. In such a situation, the ball quickly reaches a terminal velocity, which is proportional to one over the frictional force. (In the same way, a parachutist reaches a terminal velocity due to air friction, even though he is still falling.) That makes the kinetic energy of the field, which alters the value of w from -1, proportional to 1 over the expansion rate of the universe squared (H^2). This allows us to calculate an approximate formula for the difference between w and -1 as a function of time: $[w(t) - (-1)] \approx [w_0 - (-1)] \times H_0^2/H^2(t)$, where w_0 is the value of w at present epoch, H_0 is the value of the Hubble constant at the present epoch, and $H(t)$ is the value of the Hubble constant at time t. If w is close to -1, we can solve Einstein's field equations to find $H(t)$. The formula Zack and I developed uses this estimate of $H(t)$ to figure out the time evolution of $w(t)$, which can then be plugged back into Einstein's field equations to determine an improved estimate for $H(t)$; this, in turn, is used to improve the estimate of $w(t)$ still further, and so on. After a few iterations, we converge on an exact solution that gives $H(t)$ and $w(t)$.

In the early universe, when the expansion rate is high, the frictional force is high and the terminal velocity of the field must be low—essentially zero. The friction is so great that even though we are on a hill, the ball is essentially stuck in place—the field is not changing—as though we were sitting at the bottom of a valley, and therefore $w = -1$. Thus, at early times we expect w to be very nearly equal to -1. But at later times, the expansion rate of the universe becomes slower and the frictional force becomes less—so the ball can begin to move. The velocity of the field generates a positive kinetic energy and a positive pressure, which, added to the vacuum energy, give $w > -1$. The simple formula Slepian and I developed for this evolution of w and H over time can be compared with the observations.

What does the future hold if $w > -1$? The landscape may be simpler than we have supposed. If supersymmetry (basic to string theory) were unbroken, then (as Bruno Zumino showed in 1975) the vacuum effects of particles and their supersymmetric partners would cancel out exactly, ultimately leaving a vacuum energy of exactly zero—sea level. However, supersymmetry *is* broken today, the supersymmetric partners having acquired large masses relative to normal particles, so the situation is more complicated, and we must also explicitly introduce gravity into the theory. Steven Weinberg (1987) noted that "it is conceivable that in supergravity the property of having zero effective cosmological constant survives to low energy without any symmetry to protect it, but this would run counter to all our experience in physics." Yet we do live with a vacuum energy density close to zero, and we may yet find a principle that selects the zero-density vacuum as the ultimate destination. Thus, the vacuum energy may be low today simply because some low-energy fields are still running downhill toward sea level. In this case, the vacuum energy will continue to decrease, and the time it takes for the universe to double in size will lengthen. Eventually, the low-grade inflation we see today will end; low-mass particles (axions) will be created and then thin out themselves. The vacuum energy will run down to sea level. The expansion will no longer be accelerated but will become slightly decelerated, due to the low-mass particles. As these particles thin out, spacetime will become nearly flat and empty. There will be no event horizons and no Gibbons and Hawking radiation, and the temperature of the universe will go asymptotically to zero. There will be no Boltzmann brains to worry about.

In this case, according to an argument by Freeman Dyson, intelligent life might continue on forever, in principle, by thinking ever more slowly, using less and less energy, and dumping waste heat into the ever-cooling cosmic microwave background. Intelligent life in the far future would have to become a thin cloud composed of electrons and positrons—the only ordinary matter particles left. These particles would have to produce large, complex structures without annihilating each other—difficult to accomplish in practice but possible in principle.

If the Higgs vacuum is stable, as most people expect, this slowed, smooth expansion would continue for a long time.[3] However, a new problem would interrupt this seemingly peaceful future. Other bubble universes created in the early inflationary history of the multiverse (as shown in Figure 5.4) will eventually collide with ours at late times and you will start to see them. Tom Statler and I (Gott and Statler 1984) considered this scenario. Since tunneling to form a bubble universe in the inflationary epoch is a rare event, it will take a long time before we are likely to see another bubble crashing into our own. Using an estimate of 1 per $10^{1,000}$ causal horizon volumes per doubling time for the rarity of bubble formation during inflation and assuming the universe had about

1,000 doublings in size during inflation *within* our bubble, the first other bubble universe colliding with us might likely appear in about $10^{1,300}$ years. If this was a universe having properties similar to our own, according to calculations by Wainwright and colleagues (2014), this would appear first as a hot spot in the cosmic microwave background. This radiation could burn you up if you were intelligent life made of a fragile, super-cold cloud of positrons and electrons. In standard dark energy, $w = -1$, the causal horizons associated with the continuing exponential expansion help shield us from attack by these external bubbles.

After the expanding sibling universe bubble wall collides with ours, things could get hotter and messy. Things would eventually cool back down and a slow expansion would continue. Eventually (on a time scale of 10^10^34 years, by Linde's calculations), negative-energy vacuum bubbles will form and hit you. In the slow-roll case, the expansion in the end is not exponential but only linear or slower. This means that the negative-energy bubbles will percolate, colliding with each other and blocking all ways to the future. Space is not expanding fast enough to keep them from hitting each other and filling the volume of space completely. Rather than an eternally fizzing champagne, the trip to the future would be more like a voyage upward in a glass of beer, where one eventually finds a froth of colliding bubbles (the head on the beer) blocking your way. Once you enter a negative-energy bubble, the insides of the bubble begin sharply contracting, ending with a Big Crunch singularity. The bubbles are hitting each other, filling up space completely, and our universe is ultimately ending in a series of Big Crunch singularities about 10^10^34 years from now.

Here we enter a realm where we would have to understand quantum gravity—how to unite general relativity and quantum mechanics—to truly understand what, if anything would happen next. Although our universe would apparently end, the multiverse as a whole would continue on, extending far beyond our view, hidden by event horizons. The inflationary state that started our universe would continue on at these extreme distances, continuing to expand and spawn sibling bubble universes like ours forever. The distant multiverse would look like an eternally inflating, very dense, eternally fizzing champagne. But we would never see these distant realms. The space between us and them is expanding so fast that light from them can never reach us.

What we as astronomers can do now is to measure the value of w today and use it to forecast how long the current period of accelerated exponential expansion propelled by dark energy is likely to last before the expansion slows. The inflationary epoch in the early universe went for at least 87 doublings, expanding the universe by a factor of at least 10^{26} and perhaps much more. If the current low-grade inflation (the accelerated expansion) we are seeing now is similar to the early inflation in this regard, then the universe would continue with its accelerated expansion for at least another trillion years and expand by at least a further factor of 10^{26} before continuing its expansion at a linear or slower rate. A precise measurement of w will tell us how long the exponential expansion will last and give us a road map for the future of the universe. Advances in understanding the Higgs field and the field responsible for inflation may ultimately be required to predict what will happen after the current exponential expansion has ended in the future.

Phantom Energy ($w < -1$)

What if $w < -1$? That is, what if w is even *more negative* than -1? Say we find a current value of $w_0 = -1.05$, for example? This would occur if we had something called *phantom energy*, as proposed by Robert Caldwell and his colleagues Mark Kamionkowski and Nevin Weinberg of Dartmouth in 2002 and 2003. In phantom energy, the kinetic energy of the field that produces the vacuum energy is negative. It's like a rolling ball having negative kinetic energy. In this case, to conserve energy, the ball will roll *up* the hill. As it gains potential energy by rising in altitude, its kinetic energy becomes more and more negative. This sounds unphysical to me—but not to a number of physicists who take it seriously. Another possibly unphysical aspect of phantom energy is that it violates the weak energy condition. That is, some observers traveling rapidly in rocket ships would actually observe the vacuum energy density to be negative. (This is due to the way pressure and energy density relate in special relativity as seen by observers traveling in rocket ships at high speed.) That would be strange. But we do know of some quantum vacuum states that violate the weak energy condition (such as the Casimir vacuum state between parallel metal plates and the quantum vacuum state around black holes that allows them to evaporate with time), so this peculiarity does not seem to be a real impediment. By the way, the slow-roll dark energy formula Zack Slepian and I developed can also be used to chart phantom energy as we roll slowly *up* the hill; it's just that, in this case, the values of $[w(t) - (-1)]$ and $[w_0 - (-1)]$ would both be negative.

If we have phantom energy, our universe will be rolling up a hill, toward higher and higher values of vacuum energy density, ultimately reaching the Planck density. The gravitational repulsive effects of the dark energy will become greater and greater. The time for the universe to double in size will get shorter and shorter. It is a runaway effect. In a finite time, perhaps as short as 150 billion years in the future, the vacuum energy density could run all the way up to the Planck density (5×10^{93} grams per cubic centimeter) and form a singularity, called a *Big Rip singularity*. As the vacuum energy density grew, its gravitationally repulsive effects would start to tear up galaxies, then solar systems. When the vacuum energy density had risen to a level of order 5 grams per cubic centimeter (equal to the density of Earth), any planets like Earth that might be left would be ripped apart. Eventually, atoms and then atomic nuclei would be torn apart as the vacuum energy density proceeded upward toward the Planck density. At the Big Rip, *everything* gets ripped apart. This is a remarkably catastrophic end for the universe and one that Caldwell and his colleagues (assuming $w_0 = -1.5$) even thought could occur in the next 22 billion years. If we observe w_0 to be more negative than -1, this is what the future might look like. You might also roll up to the top of a hill and get stuck there. This could leave you with a very fast doubling time and a repulsive gravitational effect still strong enough to tear you apart—equally catastrophic in practice.

Baryon Acoustic Oscillations—Using the Cosmic Web to Measure Dark Energy

After 9 years of taking data, the WMAP satellite was able to measure the distance to the cosmic microwave background quite accurately and then combine this with distances to galaxies at intermediate redshifts from other methods to chart the expansion history of the universe. Using a simple model where w was assumed to be constant with time, the best fit to all the data was $w = -1.073 \pm 0.09$. Remarkably, within the errors, this was consistent with the simple value of $w = -1$ (approximating Einstein's cosmological constant), which corresponds to the model where we are sitting at the bottom of a valley. Data from the Planck satellite yielded a similar estimate. Einstein should be proud. While strongly supporting the general idea that dark energy represents a vacuum state with positive energy and negative pressure, these observations were not yet able to distinguish models in which we are sitting at the bottom of a valley from those in which we are slowly rolling down (or up) a hill. In those latter cases, the value of w would be close to -1 but slightly above (or below) it. More accurate observations were called for.

One way to measure w with greater accuracy is simply to keep measuring the distances to more and more supernovae of type Ia. This work will surely be done, but the cosmic web offers another method. The cosmic web stretches with time as the universe expands; if we can measure the sizes of structures in the cosmic web as a function of time, we can chart the expansion history of the universe to see how fast it is accelerating and thereby measure w.

Daniel Eisenstein, a Princeton valedictorian and now professor at Harvard, devised an ingenious method. He noticed that the cosmic web is imprinted with a special scale due to sound waves in the early universe. Most of the mass in ordinary matter is in particles called *baryons* (protons and neutrons). Before recombination, most of the mass in ordinary matter is in free protons and the electrons are separate. The electrically charged electrons are electromagnetically coupled to the cosmic microwave background radiation and drag the oppositely electrically charged protons with them. This couples the ordinary matter to the radiation, as we have discussed. Because of the radiation pressure of photons, radiation does not like to be squeezed, and so there is a restoring force pushing back on any fluctuation in the baryons (mainly protons) that carry most of the mass of normal matter. If you push on something (the radiation) and it pushes back, you get oscillations—these are oscillating sound waves. Perturbations in the baryons (protons and neutrons) create these sound waves (hence the name *baryon acoustic oscillations*). Having extra baryons and dark matter at a location is like dropping a rock in a pool: a circular sound-wave ripple expands out from the perturbation. Imagine the water freezing; a circular ripple would be left in the ice, and excess baryons would be left there in the ripple. Once recombination occurs, the sound waves end, and the ripple is frozen in. Then the ripple can grow via gravity. Therefore, if we look at any galaxy in the cosmic web today (like the location of the rock in the pool) and measure its covariance function (the excess probability of finding another neighboring galaxy as a function of radius), we should see this ripple. In other words, if we live on a galaxy, we should see a slight excess of other galaxies above the average trend when you look at a particular radius corresponding to the ripple.

We can figure out the exact distance to this ripple using our cosmological model. We know the fluctuations we expect from inflation, and we understand the physics of the sound waves, so we can calculate the radius of the ripple at recombination. It depends on the current value of the Hubble constant, the value of the baryon density, and the value of the total matter density (dark matter plus baryons), but we can estimate the relevant combination of these parameters from the microwave background itself. Furthermore, we can actually see the ripple in the microwave background and figure out the ripple radius at recombination: it's about 0.46 million light-years. Multiplying by the factor by which the universe has expanded from recombination until today (1,090), we get the radius of the ripple today, which is about 500 million light-years (see Color Plate 15).

To look for the ripple in the cosmic web, Eisenstein decided to measure the covariance function of the luminous red galaxies (LRGs) in the Sloan Digital Sky Survey. He picked these galaxies because they are massive elliptical galaxies that are seen only in the densest regions. They are often found in large clusters and are biased in their formation locations toward higher-density regions. Such galaxies were known to have covariance functions of higher amplitude than those for normal galaxies (making the ripple easier to detect). Furthermore, the Sloan Survey could see the LRGs out to larger distances because they were so bright—therefore, a larger volume of the universe could be surveyed using them. Eisenstein and his team found the ripple as a bump in the covariance function at the expected place (see Color Plate 15, bottom). The data are shown as black dots with error bars. The variable s is the distance between galaxies. At small scales the covariance function falls off like $s^{-1.8}$

[equivalent to $r^{-1.8}$]. But at large scales one can clearly see the baryon acoustic bump at about 105 h^{-1} megaparsec = 510 million light-years (with $H_0 = 67$ and, therefore, $h = 0.67$)—shown in close-up at right. The upper three curves (green, red, and blue) represent plausible values of $\Omega_{\text{matter}} = 0.27$, 0.29, and 0.31. The smooth magenta curve is a pure cold dark matter model without the baryon acoustic bump, which is a poor fit. This was the first detection of the baryon acoustic bump.

This could now be used as a tool to measure the expansion of the universe. Say one was looking at a sample of LRG galaxies that had a redshift of 0.35 due to the expansion of the universe. A distant galaxy at that redshift will be surrounded by a spherical shell of extra galaxies that represent the ripple. It's like looking at a spherical Christmas tree ornament in the distance. If we know the actual physical size of the Christmas tree ornament and we know its angular size as it appears to us (how big it looks in the sky), we can estimate its distance. We do this in everyday life: if an object of known size looks smaller, we deduce that it must be further away. Eisenstein could measure the angular size of that shell in the sky, and knowing the shell's actual size, he could figure out the distance to the galaxy at the center of the shell. Of course, Eisenstein and his colleagues (in 2005) averaged over many galaxies. This allowed them to measure statistically the distance of galaxies at a redshift of 0.35. By repeating this procedure at different redshifts, we can plot how distances are increasing with redshift and figure out the expansion history of the universe. If we plot all galaxies at the distances we expect from their redshift, we get a second check by examining the shape of Christmas tree ornament (the ripple); if our cosmological model is correct, it will be spherical and not elongated or squashed in the direction of Earth. This shape information allows us to independently measure the Hubble constant as a function of time as a check on our expansion history calculations.

Once he had this good idea, Eisenstein set out to form a larger group to amass more data—big science at its best. The Sloan Survey was coming to an end, so Eisenstein gathered people from the Sloan group to propose an extension of the survey to larger redshifts using LRG galaxies only: it would be called the Sloan Digital Sky Survey-III, or SDSS-III for short. The LRG galaxies would be identified by color from the original Sloan digital sky images, and then a new, more sensitive multifiber spectrometer would be used with the SDSS telescope to measure the redshifts of these galaxies. With the redshifts of all the LRG galaxies in hand, one could make a 3D map of their distribution, measure the covariance function, find the ripple, and measure the average distance to galaxies at a redshift of 0.6, for example. Over a hundred people have worked on this Baryon Oscillation Spectroscopic Survey (BOSS) project.

Nonlinear galaxy clustering can smear out the ripple a little, but with a 3D map of galaxy locations, one can fairly accurately figure out which direction gravity has been pulling the galaxies all these years since recombination. Then on the computer one can try *reversing* that history, moving them back closer to their locations in the initial conditions. This can sharpen the ripple somewhat, making the distance measurement more accurate. Eisenstein estimated that using this technique, the BOSS survey could calculate the average distance to redshift 0.6 galaxies to an accuracy of ±1.1%. The more galaxies that are measured, the more accurate will be the determination of their average distance. Measure distances as a function of redshift and, like the supernova teams, one can chart the expansion history of the universe. This is an exciting use of the cosmic web to measure dark energy, using a ripple imprinted on the cosmic web to tell us how the web, and therefore the universe, is expanding with time.

The BOSS team has applied the same BAO method to hydrogen clouds at a redshift of 2.4 to determine their distance. Quasars have absorption lines in their spectra, which are due to hydrogen clouds lying between them and us. We know hydrogen absorbs at a wavelength of 121.6 nanometers; by measuring the observed absorption-line wavelengths, we can determine the redshifts of each of these clouds. We also know the line of sight on which they are seen. We can thus make a 3D map of the positions of these hydrogen clouds in space, measure their covariance function, and find the ripple. This allows us to use the BAO ripple to measure the average distance to objects of redshift 2.4.

The Planck satellite has measured the acoustic ripple radius in the microwave background and has found it to be 1.194°. The accuracy of this measurement is 0.1%, allowing us to make a very accurate estimate of the distance to the cosmic microwave background at a redshift of 1,089, when the universe was 1,090 times smaller than it is today. Putting all these measurements together with distances to supernovae of type Ia at various redshifts, we can get an accurate history of the expansion of the universe.

In November 2014, the SDSS survey announced its long-awaited results using the BAO technique, giving us the most accurate measurement of w so far. This was truly a team effort involving over a hundred astronomers, led by Daniel Eisenstein. I was one of the many coauthors of this paper (Aubourg et al. 2014).

The first thing the paper did was to make a very accurate estimate of H_0 by using the BAO distance measurements to set the overall scale, while using supernovae measurements to chart the detailed changes in the Hubble constant with time. In this way, the BAO distances could replace distances based on Cepheid variable stars, which have always been a significant source of uncertainty. The answer was

$$H_0 = 67.3 \pm 1.1 \text{ kilometers/second/megaparsec.}$$

This agrees extraordinarily well with the estimates by the Planck team of 67 ± 1 and 67.8 ± 0.9. I did not work on this part of the SDSS project. When I first heard the result from David Weinberg, I was quite excited, because not only did it agree with the Planck value so well, but it also agreed with the median value of 67 my colleagues and I had found using median statistics (Gott, Vogeley, Podariu, and Ratra 2001).

Results on w, using the slow-roll formula that Zack Slepian and I developed, and published in collaboration with Joel Zinn, are given in Aubourg and colleagues (2014). The formula assumes that w will evolve in time, according to the physics of a vacuum field either rolling slowly downhill to lower energies (slow-roll dark energy) or slowly rolling uphill to higher energies (slow-roll phantom energy). The formula works equally well for both models. The result, which combines the Sloan Digital Sky Survey data on baryon acoustic oscillations with supernova data and Planck cosmic microwave background data, is

$$w_0 = -0.95 \pm 0.07 \text{ (68\% confidence).}$$

The best-fit value of $w_0 = -0.95$ corresponds to slow-roll dark energy. For the first time, experimental results favored slow-roll dark energy ($w > -1$) over phantom energy ($w < -1$). Recall that the WMAP results suggested $w = -1.073 \pm 0.09$ (68% confidence). The error bars are smaller, and the results have switched to a value on the other side of -1.

As the errors have shrunk, the estimates have inched closer to the value $w = -1$. If $w = -1$ is the correct model, that is exactly what one would expect to happen. Subsequent measurements would move closer and closer to $w = -1$, with that magic value always being included within the observational errors. It would not be surprising to see the value randomly oscillate between slow-roll dark energy and phantom energy as one approached $w = -1$ with smaller and smaller errors.

Indeed, the Planck Collaboration (2015b) has just added new data to this based on weak gravitational lensing. The amount of distortion seen in background galaxies due to weak gravitational lensing depends on the distances of the background galaxies behind the lens and can thus be used to relate distances to redshifts, setting additional constraints on the expansion history of the universe. Using all this combined data: BAOs, supernova, the Hubble constant, and weak lensing, again fitting with the Slepian–Gott–Zinn formula, they find

$$w_0 = -1.008 \pm 0.068 \text{ (68\% confidence).}$$

That's as close to -1 as one could hope to get with current data. The difference between the two most recent results is 0.058, comparable to the observational error of 0.068. Since the $w = -1$ model lies at the razor's edge between slow-roll dark energy and phantom energy, if we continue to find results consistent with $w = -1$, but oscillating between slow-roll dark energy and phantom energy, those two other models will never be completely ruled out. But if we improve our observations and reduce our errors by factors of ten or even a hundred and they still are centered on $w = -1$, people may well see the handwriting on the wall and pronounce the $w = -1$ model triumphant.

Topology of the Cosmic Web

Meanwhile, Changbom Park had an idea of how we could use our genus statistic to chart the expansion of the universe. The genus counts structures in the cosmic web, and as the universe expands, the web expands and the number of structures per cubic volume goes down. Basically, the initial conditions in the cosmic microwave background tell us how many structures to expect per unit volume, and multiplying that number by the counts we observe tells us the physical volume of our survey region. Knowing the volume of our survey allows us to estimate its radius—out to its most distant galaxies at redshift 0.6. That enables us to make an independent estimate of the distance to redshift 0.6 galaxies.

Park and Juhan Kim wrote an important paper (Park and Kim 2010) on this technique and showed how nonlinear clustering effects occurring since the initial conditions were small and how our large computer simulations (which included these effects) could be used to correct for them.

My undergraduate thesis student Rob Speare and I teamed with Park and Kim to analyze the accuracy of this method using our Horizon Run computer simulations. We found that in a survey like the BOSS survey, one should be able to use this technique to measure the distance to redshift 0.6 galaxies to an accuracy of ±1.7%. That is almost as good as the BAO technique. We will be using the same LRG galaxies that the BOSS project will have already obtained. We expect to use smoothing lengths of 78, 102, and 165 million light-years, enabling us to study structure on scales different from those used in the BAO method (which uses the ripple at 500 million light-years). Therefore, our estimates should be independent. If we combine our measurement with the BAO measurement, the error in the distance to redshift 0.6 galaxies could in principle be lowered to ±0.92%. In this way, we hope to provide a significant, complementary check on the more accurate BAO method. So far, my undergraduate thesis student Prachi Parihar has measured the genus from the partially completed BOSS survey, as shown in Color Plate 12. She found a count consistent with the standard cosmology with $w_0 = -1$ within the errors.

What Will Astronomers in the Far Future See?

If $w = -1$, any astronomers from the Milky Way who might be around a trillion years from now would have rather a dull time of it, as once pointed out by Lawrence Krauss. The other galaxies would have redshifted so much that they would have effectively faded from view. The cosmic web would have stretched to vastly larger size, but most of it would remain forever hidden from us, forever beyond our event horizon. As time went on, we would find no new structures from the cosmic web coming into view. Even if we waited till the infinite future, we would see a total number of structures in the cosmic web only a factor of 2.36 times larger than the number of structures we can see today. Just a trillion years from now, the cosmic microwave background radiation will have redshifted to the point of being unobservable by present technology, and the Gibbons and Hawking radiation would also be too weak to observe given present technology. Future observers using their telescopes would see only our own galaxy, grown somewhat more massive from adding the masses of the other galaxies in our Local Group, which, like Andromeda, will have merged with it through collisions. It might be hard to tell that the universe was expanding, because one could not readily see in the visible wavelengths any tracer galaxies to mark the expansion of space. It would look like we were a lone galaxy in an empty universe. But a few clues to a more crowded past would be left: the distribution of elements would leave evidence for a Big Bang at the beginning, and any star escaping from our galaxy would accelerate as it moved away from us—tracing the effects of dark energy. If we waited long enough, a bubble of negative-energy vacuum state would eventually hit us, ending our adventures. This scenario would be the most boring for future astronomers.

If $w > -1$, the density of dark energy would decrease with time (slow-roll dark energy). As an example, if $w_0 = -0.95$ (the best-fit value from the Sloan-III BAO observations and supernovae observations), the accelerating exponential expansion of the universe we are seeing today should continue for roughly 120 billion years. In the near future the universe will be doubling in size every 12.2 billion years; that will slow to doubling in size every 51 billion years as the vacuum state slowly rolls downhill to lower energies. Overall, during this process, it will double in size 6 times. Then, about 120 billion years from now, the field will begin to oscillate, dumping the energy in the vacuum field into particles (probably ultralight axions more than 30 orders of magnitude lighter than neutrinos). At this point, a slower, decelerated expansion would take over. Galaxies that have said goodbye to us, disappearing behind a causal horizon during the dark energy's accelerated exponential expansion phase, will begin to slow down and come back into view, saying hello again. If the Higgs vacuum is somehow stabilized by interaction with a massive axion field, as many people expect, future astronomers in our universe may last undisturbed until perhaps $10^{1,300}$ years from now, when another expanding sibling bubble universe formed in the inflating multiverse collides with us. By that time, the deceleration of our universe could have brought into view vast, far reaches of our own universe that we have not seen yet. As our causal horizon expands outward during this decelerating phase, we might ultimately be able to see roughly $10^{1,800}$ times as many structures in the cosmic web as we can observe today! All these would be horribly redshifted by a factor of $10^{1,000}$. Even the highest-energy gamma ray sources would have their radiation redshifted into ultralong radio waves. Any future "astronomers" at this time would be sparse intelligent clouds of positrons and electrons trying to measure slowly changing electric and magnetic fields of minuscule magnitude. Detection would be a challenge for even the most advanced civilization, but at least there is a long time (perhaps $10^{1,300}$ years) to master the art! If these radio photons could be detected, the view would be magnificent, showing us $10^{1,800}$ times as much of the cosmic web as we can see today. (For the view today, see Color Plate 16.) This is the most exciting scenario for future astronomers.[4]

Eventually, we expect negative-vacuum-energy bubbles to form, collide, and percolate, ending our universe in a series of Big Crunch singularities. The universe would end in a froth of bubbles, like that topping off a glass of beer.

If $w < -1$, we would have phantom energy, and future astronomers would be torn apart by the increasing dark energy as they approached the Big Rip. As an example, if $w_0 = -1.008$ (the best fit value from the Planck Collaboration (2015b)), the Big Rip will occur 1.5 trillion years from now. If $w_0 = -1.076$ (at the lower limit of Planck's 68% confidence limits), the Big Rip could occur in as little as another 150 billion years.

Which of these futures awaits us? Eternally fizzing champagne, a heady beer, or a Big Rip singularity?

To determine what will happen after the current epoch of exponential expansion ends will require some breakthroughs in elementary particle physics—such as finding the long-hoped-for "theory of everything" perhaps growing out of M theory. But as for what will happen in the next trillion years, *that* is something we may fairly confidently extrapolate by simply measuring w_0 more accurately. Knowing that value will tell us if (in the next trillion years) the universe will continue its current rate of exponential expansion, doubling in size every 12.2 billion years ($w_0 = -1$), if it will slow down (slow-roll dark energy, $w_0 > -1$), or if it will speed up (phantom energy, $w_0 < -1$).

Fortunately, there are dramatic proposals in the works for increasing the accuracy of the measurement of w_0. When Lyman Spitzer (my thesis adviser) developed the idea for the Hubble Space Telescope, it was built with a diameter of 2.4 meters. It has revolutionized our astronomical knowledge, and any amateur astronomer is familiar with many of the famous pictures it has taken. It had that particular diameter of 2.4 meters because that was the diameter of the spy satellites in use at that time. It was built on the same basic design, simply looking up instead of down. Now in this epoch of budget cuts, the National Security Agency has given astronomers two 2.4-meter spy satellite telescopes that it no longer needs! What an unexpected windfall this is for basic research. NASA will deploy one of them in the quest to measure w. Supernovae of type Ia will be observed, and the BAO method will be used to study the cosmic web out to large redshifts, and measurements of weak gravitational lensing will also be made. This satellite will be called Wide Field Infrared Survey Telescope (WFIRST). It offers the prospect of improving current error bars on w_0 by about a factor of about 14 (Spergel et al. 2013). Europe is planning a complementary satellite named EUCLID. Many efforts are planned for ground-based astronomy as well. The 8.4-meter-diameter Large Synoptic Survey Telescope (LSST) located on Cerro Pachón in northern Chile will survey the southern-hemisphere sky. Another project, originally called Big BOSS (so-named because it includes fainter galaxies and goes deeper than the BOSS survey), will use the 4-meter-diameter Kitt Peak telescope to survey the cosmic web. This project has now been renamed Dark Energy Spectrographic Instrument (DESI).

In just the past century we have learned that planet Earth is just a tiny speck in a vast cosmos. We have also discovered the architecture of the universe—billions of galaxies linked in a giant cosmic web. As the fossilized remnants of quantum fluctuations made in the first 10^{-35} seconds of our universe, the cosmic web is the oldest thing we can see. It may also provide clues as to the nature of the potentially vast stretches of time that loom ahead of us in the future.

Notes

Preface

1. To account for the uniformity of the universe on large scales and its enormous size, the theory of inflation proposes that the early universe underwent a period of super-fast expansion, where it repeatedly doubled in size. This inflationary epoch explains how the Big Bang expansion of the universe got started.

Chapter 1: Hubble Discovers the Universe

1. Here is what Einstein's equation looks like: $R_{\mu\nu} - \frac{1}{2}g_{\mu\nu}R = 8\pi T_{\mu\nu}$. I just wanted you to see it! Needless to say, it requires a lot of unpacking of mathematical terms to understand it in detail, but putting it simply, the "stuff" of the universe (matter, mass energy, pressure, etc.) causes space and time to curve. The right-hand side of the equation uses the stress-energy tensor $T_{\mu\nu}$ to describe the mass-energy density and pressure of stuff, plus the momentum flux and stress associated with stuff at a point. The left-hand side of the equation tells how space and time are curved at that point. The equation as a whole describes exactly *how* the stuff of the universe (matter, radiation) causes space and time to curve.

The terms in the equation are tensors—mathematical objects that can be translated from one coordinate system to another. A vector is a tensor of rank 1. Vectors are used to show velocities; a velocity has a particular magnitude and a particular direction. It has components in each of the different directions. A vector on a two-dimensional surface has two components; wind on the surface of Earth, for example, has a North-South component and an East-West component. If one changes coordinate systems (say to Magnetic North), the components of the vector will change in a particular way, specified by the equations relating the two coordinate systems. The stress-energy tensor $T_{\mu\nu}$ is a tensor of rank two that has $4 \times 4 = 16$ components in four-dimensional spacetime. In a local coordinate system where the observer is at rest: T_{tt} = mass-energy density, T_{xx} = pressure in the x-direction, T_{yy} = pressure in the y-direction, T_{zz} = pressure in the z direction. $T_{tx} = T_{xt}$, $T_{ty} = T_{yt}$, $T_{tz} = T_{zt}$ represent components of momentum flux in the x-, y-, and z-directions, and $T_{xy} = T_{yx}$, $T_{xz} = T_{zx}$, $T_{zy} = T_{yz}$ represent components of stress in different directions. Importantly, pressure gravitates as well as energy density.

The laws of tensor arithmetic tell us how the components of the stress-energy tensor change when we change coordinate systems. The term $R_{\mu\nu}$ is a tensor derived by summing components of the even more complicated Riemann curvature tensor $R_{\mu\nu\alpha\beta}$ (which has $4 \times 4 \times 4 \times 4 = 256$ components describing how spacetime is curved at a particular point). The term $-\frac{1}{2}g_{\mu\nu}R$ is composed of the metric tensor $g_{\mu\nu}$ multiplied by a quantity R derived by summing components of $R_{\mu\nu}$. The metric tensor measures separations between nearby events.

It took Einstein 8 years of hard work to derive this equation; it is one of the triumphs of human thought. I'm showing it to you so you might appreciate its complexity and yet its ultimate elegance and simplicity. It embodies a simple idea, that the stuff in the universe causes spacetime to curve—and that is what is responsible for gravity. Amazing.

2. To produce a static universe, Einstein added a new term to his field equation, making it read: $R_{\mu\nu} - \frac{1}{2}g_{\mu\nu}R + \Lambda g_{\mu\nu} = 8\pi T_{\mu\nu}$. The new term $\Lambda g_{\mu\nu}$ was called the *cosmological constant* term. It is the metric tensor multiplied by a constant Λ (capital lambda). R is a measure of the overall curvature and can vary throughout spacetime, but Λ is a universal constant that is constant throughout spacetime. As pointed out by Lemaître, the term may be moved to the opposite side of the equation: $R_{\mu\nu} - \frac{1}{2}g_{\mu\nu}R = 8\pi T_{\mu\nu} - \Lambda g_{\mu\nu}$, where it may be interpreted as a vacuum energy density: $R_{\mu\nu} - \frac{1}{2}g_{\mu\nu}R = 8\pi[T_{\mu\nu} + (T_{\mu\nu})_{\text{vac}}]$, with $(T_{\mu\nu})_{\text{vac}} = -\Lambda g_{\mu\nu}/8\pi$. The structure of the metric tensor (which locally can be written as $ds^2 = -dt^2 + dx^2 + dy^2 + dz^2$ or $g_{tt} = -1$, $g_{xx} = 1$, $g_{yy} = 1$, $g_{zz} = 1$) with opposite signs between the time and space components ensures that when multiplied by $-\Lambda/8\pi$, it gives opposite signs for the vacuum energy and the vacuum pressure: a positive vacuum energy $(T_{tt})_{\text{vac}}$ must be accompanied by a negative vacuum pressure in the x-, y-, and z-directions: $(T_{xx})_{\text{vac}}$, $(T_{yy})_{\text{vac}}$, $(T_{zz})_{\text{vac}}$. This negative pressure has a gravitationally repulsive effect that is larger than the gravitational attractive effect of the energy density in the vacuum by a factor of 3 because there are 3 dimensions of space and only 1 dimension of time. Thus, the overall effect of the vacuum energy is repulsive, and can balance the gravitational attraction of the galaxies for each other to produce a static universe.

3. Einstein introduced the cosmological constant to allow a static cosmological model. Interestingly, if Einstein had been a Flatlander (living in a universe with only two spatial dimensions instead of three), he wouldn't have needed to invent the cosmological constant. Mark Alpert and I (Gott and Alpert 1984) showed that Einstein's field equation (without a cosmological constant) in this case had a static cosmological solution: a universe of uniform density, shaped like the surface of a sphere whose radius did not change with time (Sphereland). Point masses in Flatland do not attract each other gravitationally and so a static solution is possible. A Flatlander Einstein could have gotten the static geometry he wanted without having to introduce the cosmological constant.

Chapter 3: How Clusters Form and Grow

1. This distance to the cosmic microwave background of 13.8 billion light-years is known as the *lookback-time distance*. The photons we see from the cosmic microwave background were last scattered by electrons 13.8 billion years ago and have been traveling through curved spacetime for 13.8 billion years; therefore, they have traveled 13.8 billion light-years. This is the distance from us back to those electrons off which the photons last scattered. But we are seeing these electrons where they were in the past, 13.8 billion years ago, just 380,000 years after the Big Bang. Where are those same electrons

now? They have been expanding outward with the expansion of the universe and are, at the current epoch, about 46 billion light-years away. This is called the *comoving distance.* It marks the radius of the visible universe at the present epoch. How did those electrons that last scattered the cosmic microwave background photons get out to a distance of 46 billion light-years from us in just 13.8 billion years? This occurs because the space between us and them has been expanding faster than the speed of light, something allowed by Einstein's theory of General Relativity.

Usually we shall be using the lookback-time distance when we are talking about the distance to a galaxy. But in the Map of the Universe (Chapter 9), which presents a snapshot of the universe at the current epoch—specifically, where objects were when the map was made on August 23, 2003—we will use the comoving distance. In that map we are interested in what distances the objects will have attained by the current epoch, not where they were in the past when they emitted the light we are seeing now. We also use comoving distance in most maps of large-scale structure, including Figure 9.6, where it is mentioned explicitly. In these pictures we are interested in making a snapshot of the universe at the current epoch.

2. Peebles' calculation went as follows. The collapse time of a cluster T_c is proportional to $(\delta\rho/\rho)^{-3/2}$ in the initial conditions. In a cluster today, ρ is proportional to $1/T_c^2$. So ρ today is proportional to $(\delta\rho/\rho)^3$ in the initial conditions or proportional to $M^{-3/2}$ for Poisson initial conditions. And, of course, today—by definition—$\rho \sim M/r^3$. Thus, $M^{-3/2}$ is proportional to M/r^3, making $M^{5/2}$ proportional to r^3 and M proportional to $r^{6/5}$, so ρ $(\sim M/r^3)$ surrounding a galaxy in the hierarchy out to a radius r is proportional to $M^{-3/2}$ and $r^{-1.8}$.

Chapter 5: Inflation

1. Thermal radiation gravitationally decelerates the worldlines of particles participating in the expansion twice as much as Newton could have predicted because of the gravitationally attractive effects of the radiation pressure that Einstein's theory predicts. Including this effect was essential in the Gamow, Herman, and Alpher calculations of the formation of the light elements in the first 3 minutes after the Big Bang. The fact that modern nucleosynthesis results predict abundances for these light elements in agreement with what we observe supports Einstein's conclusion that positive pressure is gravitationally attractive.

2. The vacuum state in inflation has a positive energy density and a negative pressure. These have opposite signs, because time and space have opposite signs in special relativity. The metric tensor that measures distances between events has the form $ds^2 = -dt^2 + dx^2 + dy^2 + dz^2$ in special relativity. The last two terms represent the Pythagorean theorem from Euclidean plane geometry: $ds^2 = dy^2 + dz^2$ (the square of the hypotenuse is equal to the sum of the squares of the two sides—along the coordinate directions dy and dz). The last three terms represent the Pythagorean theorem generalized to give the usual 3D result from Euclidean solid geometry. The minus sign on the dt^2 term guarantees that the separation in time [in years] (dt) and space [in light-years] ($\sqrt{[dx^2 + dy^2 + dz^2]}$) between two events connected by a light ray (where $ds^2 = 0$) are equal, ensuring that all observers will observe light traveling at the same speed through empty space—a fundamental postulate of special relativity. It is that little minus sign in front of the dt^2 term that makes all the difference between time and an ordinary

spatial dimension such as width. In general relativity, energy density is associated with the time dimension, while the pressure in the x-, y-, and z-directions is associated with the x, y, and z spatial dimensions. The vacuum state is proportional to the metric tensor and, therefore, has opposite signs for its energy density and pressure (in the x-, y-, and z-directions).

Chapter 6: A Cosmic Sponge

1. Many years later, I was asked to serve as a judge for the Westinghouse Science Talent Search and then as chair of the judges. I was chair of the judges for 14 years and saw the contest into its first year as the Intel Science Talent Search. Glenn Seaborg was a member of my judging panel. The students and I met with Presidents George Herbert Walker Bush and Bill Clinton during this time. One of our first-place winners was Jacob Lurie, who had a project in surreal numbers in mathematics. He would go on to become a professor at Harvard and win the inaugural Breakthrough Prize in Mathematics in 2014 ($3 million), as well as a MacArthur "genius grant." He is just one example of the many wonderful students who passed through the contest during those years and have gone on to great careers in science.

2. Tom Banchoff was an instructor in math at Harvard when he encouraged me to publish my paper on pseudopolyhedrons. He went on to become a professor at Brown University. He famously became a friend of Salvador Dali, who was interested in learning more about developments in higher-dimensional geometry for possible use in his paintings. Banchoff recently provided commentary on higher-dimensional geometry on the DVD of the charming 2007 animated film *Flatland: The Movie*, based on the famous 1884 novel by Edwin A. Abbott, describing what life in a world with only two spatial dimensions would be like.

3. The N-body simulations assumed periodic boundary conditions on the cubical sample. The computer used a very efficient fast Fourier algorithm to calculate the forces on the particles as if the cubical sample was a video game in which: if you exited out the top, you reentered at the bottom; if you exited out the right side, you reentered at the left; and if you exited out the front, you reentered at the back. This makes it possible for the computer to handle the maximum number of particles, in the least amount of computer time. This is standard for large N-body simulations today. For comparative purposes, in our paper we treated our observational cubical CfA sample the same way. Thus, we were effectively assuming that the universe was an infinite warehouse filled with cubical boxes that were copies of our CfA sample. We found that our median density contour, which divided the universe into a high-density half and a low-density half by volume, was spongelike and all in one connected piece. In Figure 6.8a the triangular-shaped piece of the median density contour visible at the top front of the cube facing you is connected to the rest of the median density contour at the top of the back-right face of the cube because of the periodic boundary conditions. In this periodic universe the high-density regions are all in one connected spongelike piece and the low-density regions are all in one connected, interlocking, spongelike piece. [The N-body computer simulation cubes portrayed later in the book all have periodic boundary conditions, but in subsequent observational data sets, where many structures are present and the samples have a non-cubical shape, the periodic boundary conditions were dropped and the sample was analyzed without assuming it lived in a universe with many other copies of itself.

The Voronoi honeycomb universe simulation in Figure 6.11 also has periodic boundary conditions, and yet here the median density contour is broken up into separate pieces enclosing isolated voids. In the 50%-low picture you can see one large void extending out the top of the survey and reentering at the bottom, for example. This is a Swiss cheese topology, and the periodic boundary conditions do not affect that.

4. Euler's formula may be applied to calculate the genus of spongelike polyhedrons. For example, the pseudopolyhedron having squares, 5 around a point (see Figure 6.4) consists of two planes of squares connected by a periodic array of cubical holes. It can be constructed from a repeated arrangement of 8 vertices, 20 edges, and 10 faces having an Euler characteristic $(V - E + F) = -2$. Every time we add 8 more vertices and their associated edges and faces to the structure, we add one to the genus and create a new cubical hole. The structure is infinite and so has an infinite number of holes. One can prove that calculating the genus by adding up the angle deficits (or excesses) at vertices, and calculating the genus by using the Euler characteristic $(V - E + F)$ always give the same answer.

Chapter 8: Park's Simulation of the Universe

1. We can estimate the physical size of a galaxy (from, for example, the internal velocity dispersion of its stars). Comparing its physical size with its angular size in the sky allows us to calculate its distance. That distance tells us via Hubble's law what we expect its radial velocity away from us to be. Subtracting that from its actual radial velocity gives us its individual or *peculiar* velocity in the radial direction relative to that expected due to the average Hubble expansion of the universe. This gives us only the component of the peculiar velocity in the radial direction (pointing toward Earth). This velocity is due to the component in the slope of the gravitational potential in the direction of Earth. If we integrate such slopes along an entire line of sight, we can map the gravitational potential along the entire line of sight. Repeat this along different lines of sight spread over the entire sky, and we can construct a 3D map of the gravitational potential throughout space. Then we can measure the slopes in this gravitational potential topography map in 3D, giving us the peculiar velocities of galaxies in 3D. This is the clever technique developed by Bertschinger and Dekel. It allows us to turn peculiar velocities in the radial direction only into a 3D map of the peculiar velocities showing their components in other directions as well.

Chapter 10: Spots in the Cosmic Microwave Background

1. Interestingly, in 3D the genus (number of donut holes – number of isolated regions in a density-contour surface) is equal to $-1/4\pi$ times the Gaussian curvature integrated over the contour surface, while the 2D genus (number of hot spots – number of cold spots) is equal to $1/2\pi$ times the curvature encountered driving around the 2D contour line. In both cases the genus is directly related to curvature integrated over the boundary.

2. Standard slow-roll inflation (a simple field rolling down a hill) predicts values of the non-Gaussian parameter f_{NL} from 10^{-2} to 10^{-1} (close to zero and undetectably different from zero with current data) according to calculations by Juan Maldacena and others. The COBE results set 68% confidence limits of $-1{,}500 < f_{NL} < 1{,}500$. Using the higher-resolution WMAP data in 2007, Wes Colley and I and our colleagues, using our

genus topology technique, were able to improve these limits to $-101 < f_{NL} < 107$ at the 95% confidence level. The WMAP team found $-58 < f_{NL} < 134$ with an independent analysis. After 6 more years of taking data, the WMAP team was able to narrow these limits to $-3 < f_{NL} < 77$. All these ranges are consistent with f_{NL} near zero. The Planck satellite, drawing upon a still-higher resolution map and testing for random phases with a different (bispectrum) method that uses temperature correlations found between triples of points in the map, has recently found $-3.1 < f_{NL} < 8.5$ (at 68% confidence), again consistent with f_{NL} near zero and the predictions of standard inflation.

Chapter 11: Dark Energy and the Fate of the Universe

1. I argued in a *Nature* paper in 1993 that we are not likely to find ourselves in a special position *among* intelligent observers. I described this as an application of the *Copernican principle*: that our location should not be special among those locations occupied by intelligent observers. Vilenkin calls this the *principle of mediocrity*. Same concept.

2. Paul Steinhardt and Neil Turok (2002) have argued that the "inflationary state" at the beginning of our universe was really the accelerated dark-energy expansion at the end of a previous universe. This inflation would occur at low energy (leading to a prediction of no significant gravity waves) and the universe would reheat by a collision of membranes in an 11-dimensional spacetime derived from M-theory. This scenario would also produce Gaussian random-phase initial conditions for our universe. Linde envisions the multiverse as an infinite, never-ending fractal tree of inflating universes with the branching produced by quantum fluctuations in the inflating state. In this model our universe is born as just one pocket universe on one of the branches. The question of where the original trunk came from is discussed in my book *Time Travel in Einstein's Universe*. Ideas range from quantum tunneling from nothing (where a de Sitter waist, like that at the bottom of Figure 5.4, simply pops into existence and inflates from there (Vilenkin 1982; Hartle and Hawking 1983)) to time loops allowing the multiverse to be its own mother (Gott and Li-Xin Li 1998).

3. Given the current best estimates of the mass of the Higgs boson, and the mass of the top quark, the Higgs vacuum could decay into negative-vacuum-energy bubbles in about 10^{42} years in the slow-roll dark energy model. Because the expansion is not exponential in the end but only linear or slower, the negative-energy bubbles will percolate, colliding with each other and blocking all ways to the future. Rather than an eternally fizzing champagne, the journey to the future would be more like a voyage upward in a glass of beer, where you eventually find a froth of colliding bubbles (the head on the beer) blocking your way. Once you enter a negative-energy bubble, the insides of the bubble begin sharply contracting, ending with a Big Crunch singularity. The bubbles are hitting each other, filling up space completely, and the universe is ending in a series of Big Crunch singularities. If the Higgs vacuum state is stabilized by higher-energy physics—as Arkani-Hamed thinks likely—the negative-energy bubbles would form less frequently, on a time scale of 10^10^34 years within the observable universe, perhaps, and you would likely be hit first by another expanding sibling bubble universe like ours left over from the early inflationary epoch (on a time scale of $10^{1,300}$ years), as we will discuss.

4. How much more of the cosmic web we will be able eventually to see depends critically on how many doubling times of exponential expansion in the past occur during inflation within our universe versus how many doubling times will occur in the future during the exponential expansion occurring in dark energy today. If the universe doubles in size many more times during inflation than during the dark energy–dominated epoch starting now, we will see vast new regions of the cosmic web.

A large number of doublings in size during inflation within our bubble after our universe forms shields us from impending sibling bubble universes by effectively pushing them further away. That allows us enough time to recover the galaxies in our universe we have said goodbye to during the dark energy phase and say hello to many new ones.

The calculation in the text assumed bubbles form at a rate of one per $10^{1,000}$ causal horizon volumes per doubling time during inflation in the multiverse. Quantum tunneling to form a bubble universe is a rare event because you must tunnel through a high mountain range to get from the high valley you start in to a place where you can roll down to the nearest low-altitude valley in the landscape. The numbers could, in principle, be even more dramatic. If the mountains in the landscape are high enough to produce a vacuum energy equal to the Planck density, we could have 10^7 doublings inside our bubble and a tunneling rate of 1 per (10^10^5) causal volumes per doubling time in the inflationary multiverse, for example. In that case, a factor of 10^10^5 more structures in the cosmic web than we can see now would come into view.

We won't know exactly how many doubling times our universe has inflated after its formation or how rare bubble formation really is until we have obtained a "theory of everything" based, hopefully, on some version of M-theory. Only then will we be able to know the landscape well enough around our particular valley to calculate just how many doubling times of inflation we are likely to see within our bubble and how long it will be before another bubble universe or pocket universe is likely to hit us. Suffice it to say that in the slow-roll dark energy scenario, there is the possibility for vast new regions of the distant cosmic web to come into view in the future. This does not occur in the $w_0 = -1$ or $w_0 < -1$ (phantom energy) models.

References

Aarseth, S. J., E. L. Turner, and J. Richard Gott. "*N*-body simulations of galaxy clustering. I— Initial conditions and galaxy collapse times." *Astrophysical Journal*, 228: 664, 1979.

Adler, R. J. *The Geometry of Random Fields*. Chichester: Wiley, 1981.

Aguirre, A., S. Gratton, and M. C. Johnson. "Hurdles for recent measures in eternal inflation." arXiv: hep-th/0611221v1, 2006.

Albrecht, A., and P. J. Steinhardt. "Cosmology for grand unified theories with radiatively induced symmetry breaking." *Physical Review Letters*, 48: 1220, 1982.

Alcock, C., et al. "The MACHO Project: Microlensing results from 5.7 years of Large Magellanic Cloud observations." *Astrophysical Journal*, 542: 281, 2000.

Alpher, Ralph A., and Robert Herman. "Evolution of the universe." *Nature*, 162: 774, 1948.

Arkani-Hamed, N., S. Dubovsky, L. Senatore, and G. Villadoro. "(No) Eternal inflation and precision Higgs physics." *Journal of High Energy Physics*, Issue 03, id. 075, 2008.

Aubourg, Éric, et al. "Cosmological implications of baryon acoustic oscillation (BAO) measurements." arXiv1411.1074, 2014.

Baade, W., and F. Zwicky. "On super-novae." *Proceedings of the National Academy of Sciences*, Vol. 20, p. 254, 1934.

Bahcall, N. A., and R. M. Soneira. "The spatial correlation function of rich clusters of galaxies." *Astrophysical Journal*, 270: 20, 1983.

Bardeen, James M., J. R. Bond, N. Kaiser, and A. S. Szalay. "The statistics of peaks of Gaussian random fields." *Astrophysical Journal*, 304: 15, 1986.

Bardeen, James M., Paul J. Steinhardt, and Michael S. Turner. "Spontaneous creation of almost scale-free density perturbations in an inflationary universe." *Physical Review D*, 28: 679, 1983.

Bennett, C. L. "Nine-year Wilkinson Microwave Anisotropy Probe (WMAP) observations: Final maps and results." *Astrophysical Journal Supplement*, 208: 20, 2013.

Bertschinger, E. "Cosmological detonation waves." *Astrophysical Journal*, 295: 1, 1985a.

———. The self-similar evolution of holes in an Einstein–de Sitter universe. *Astrophysical Journal Supplement*, 58: 1, 1985b.

Bertschinger, E., and A. Dekel. "Recovering the full velocity and density fields from large-scale redshift-distance samples." *Astrophysical Journal*, 336: L5, 1989.

Bhavsar, S. P., S. J. Aarseth, and J. Richard Gott. "*N*-body simulations of galaxy clustering. V—The multiplicity function." *Astrophysical Journal*, 246, 656: 1981.

BICEP2/Keck and Planck Collaborations, P.A.R. Ade, et al. A Joint Analysis of BICEP2/ *Keck Array* and *Planck* Data, 2015. (http://new.bicepkeck.org/BKP_paper_20150130 .pdf)

Blumenthal, George R., Heinz Pagels, and Joel R. Primack. "Galaxy formation by dissipationless particles heavier than neutrinos." *Nature*, 299: 37, 1982.

Boddy, Kimberly K., Sean M. Carroll, and Jason Pollack, "Why Boltzmann Brains don't fluctuate into existence from the de Sitter vacuum." arXiv:1505.02780, 2015.

Bond, J. R., L. Kofman, and D. Pogosyan. "How filaments are woven into the cosmic web." arXiv:astro-ph/9512141v1, 1995.

Brout, R., F. Englert, and E. Gunzig. "The creation of the universe as a quantum phenomenon." *Annals of Physics* (USA) 115: 78, 1978.

Caldwell, Robert R. "A phantom menace? Cosmological consequences of a dark energy component with super-negative equation of state." [arXiv: astro-ph/9908168, 1999] *Physics Letters B*, 545: 23, 2002.

Caldwell, Robert R., Mark Kamionkowski, and Nevin N. Weinberg. "Phantom energy: Dark energy with $w < -1$ causes a cosmic doomsday." *Physical Review Letters*, 91: 1301, 2003.

Canavezes, A., et al. "The topology of the IRAS Point Source Catalogue Redshift Survey." *Monthly Notices of the Royal Astronomical Society*, 297: 777, 1998.

Colless, M., et al. The 2dF Galaxy Redshift Survey: Spectra and redshifts. *Monthly Notices of the Royal Astronomical Society*, 328: 1039, 2001.

Colley, W. N., J. Richard Gott, D. Weinberg, C. Park, and A. Berlind. "Topology from the simulated Sloan Digital Sky Survey." *Astrophysical Journal*, 529: 795, 2000.

Colley, Wesley N., and J. Richard Gott. "Genus topology of the cosmic microwave background from WMAP." *Monthly Notices of the Royal Astronomical Society*, 344: 686, 2003.

Coxeter, H.S.M. "Regular skew polyhedra in three and four dimensions." *Proceedings of the London Mathematical Society*, 43 (2): 33, 1937.

Davis, Marc. "Summary: Problems and prospects for large scale structure studies." In *The Large-Scale Structures and Peculiar Motions in the Universe*, D. W. Latham and L. N. da Costa, eds., conference in Rio de Janeiro 1989, *Proceedings of the Astronomical Society of the Pacific*, p. 379, 1991.

Davis, M., G. Efstathiou, C. S. Frenk, and S.D.M. White. "The evolution of large scale structure in a universe dominated by cold dark matter." *Astrophysical Journal*, 292: 371, 1985.

Davis, M., and P.J.E. Peebles. "A survey of galaxy redshifts. V—The two-point position and velocity correlations." *Astrophysical Journal*, 267: 465, 1983.

De Lapparent, V., M. J. Geller, and J. P. Huchra. "A slice of the universe." *Astrophysical Journal*, 302: L1, 1986.

De Sitter, W. "Einstein's theory of gravitation and its astronomical consequences. Third paper." *Monthly Notices of the Royal Astronomical Society*, 78: 3, 1917.

De Vaucouleurs, Gerard. "Evidence for a local super galaxy." *Astronomical Journal*, 58: 30, 1953.

Dicke, R. H., P.J.E. Peebles, P. G. Roll, and D. T. Wilkinson. "Cosmic black-body radiation." *Astrophysical Journal*, 142: 414, 1965.

Dressler, A. *Voyage to the Great Attractor: Exploring Intergalactic Space.* New York: Knopf, 1994.

Einasto, J. *Dark Matter and Cosmic Web Story*. Hackensack, NJ: World Scientific, 2014.

Einasto, J., M. Joeveer, and E. Saar. "Structure of superclusters and supercluster formation." *Monthly Notices of the Royal Astronomical Society*, 193: 353, 1980.

Einasto, J., A. Kaasik, and E. Saar. "Dynamical evidence on massive coronas of galaxies." *Nature*, 250: 309, 1974.

Einstein, Albert. "Lens-like action of a star by the deviation of light in the gravitational field." *Science*, 84: 506, 1936.

Einstein, Albert, and W. de Sitter. "On the relation between the expansion and the mean density of the universe." *Proceedings of the National Academy of Sciences*, vol. 18, p. 213, 1932.

Eisenstein, Daniel J., D. H. Weinberg, et al. "SDSS-III: Massive spectroscopic surveys of the distant universe, the Milky Way, and extra-solar planetary systems." *Astronomical Journal*, 142: 72, 2011.

Eisenstein, Daniel J., I. Zehavi, et al. "Detection of the baryon acoustic peak in the large-scale correlation function of SDSS Luminous Red Galaxies." *Astrophysical Journal*, 633: 560, 2005.

Elias-Miró, J., et al. "Higgs mass implications on the stability of the electroweak vacuum." *Physics Letters B*, 709: 222, 2012.

Elias-Miró, J., et al. "Stabilization of the electroweak vacuum by a scalar threshold effect." *Journal of High Energy Physics*, 2012: 31, 2012.

Farhi, E., A. H. Guth, and J. Guven. "Is it possible to create a universe in the laboratory by quantum tunneling?" *Nuclear Physics B*, 339: 417, 1990.

Freeman, K. "On the disks of spiral and S0 galaxies." *Astrophysical Journal*, 160: 811, 1970.

Friedmann, A. "Über die Krümmung des Raumes." *Zeitschrift für Physik*, 10: 377, 1922.

———. "Über die Möglichkeit einer Welt mit konstanter negativer Krümmung des Raumes." *Zeitschrift für Physik*, 21: 326, 1924.

Gamow, G. "The evolution of the universe." *Nature*, 162: 680, 1948a.

———. "The origin of elements and the separation of galaxies." *Physical Review*, 74: 505, 1948b.

Garriga, J., and A. Vilenkin. "Recycling universe." *Physical Review D*, 57: 2230, 1998.

Geller, M. J., and J. P. Huchra. "Mapping the universe." *Science*, 246: 897, 1989.

Gott, J. Richard. "Pseudopolyhedrons." *The American Mathematical Monthly*, 74: 497, 1967.

———. "Are heavy halos made of low mass stars? A gravitational lens test." *Astrophysical Journal*, 243: 140, 1981.

———. "Creation of open universes from de Sitter space." *Nature*, 295: 304, 1982.

———. "Gravitational lensing effects of vacuum strings: Exact solutions." *Astrophysical Journal*, 288: 422, 1985.

———. "Conditions for the formation of bubble universes." In E. W. Kolb et al., eds., *Inner Space/Outer Space* (pp. 362–66). Chicago: University of Chicago Press, 1986.

———. "Implications of the Copernican principle for our future prospects." *Nature*, 363: 315, 1993.

———. *Time Travel in Einstein's Universe*. Boston: Houghton Mifflin, 2001.

———. "Boltzmann brains—I'd rather see than be one." arXiv:0802.0233, 2008.

Gott, J. Richard, and M. Alpert. "General relativity in a (2 + 1)-dimensional spacetime." *General Relativity and Gravitation*, 18: 243, 1984.

Gott, J. Richard, Yun-Young Choi, Changbom Park, and Juhan Kim. "Three-dimensional genus topology of Luminous Red Galaxies." *Astrophysical Journal*, 695: L45, 2009.

Gott, J. Richard, Wesley N. Colley, Chan-Gyung Park, Changbom Park, and Charles Mugnolo. "Genus topology of the cosmic microwave background from the WMAP 3-year data." *Monthly Notices of the Royal Astronomical Society*, 377: 1668, 2007.

Gott, J. Richard, and James E. Gunn. "The Coma cluster as an X-ray source: Some cosmological implications." *Astrophysical Journal*, 169: L13, 1971.

Gott, J. Richard, D. Clay Hambrick, et al. "Genus topology of structure in the Sloan Digital Sky Survey: Model testing." *Astrophysical Journal*, 675: 16, 2008.

Gott, J. Richard, Mario Jurić, et al. "A map of the universe." *Astrophysical Journal*, 624: 463, 2005.

Gott, J. Richard, and Li-Xin Li. "Can the universe create itself?" *Physical Review* D, 58: 023501, 1998.

Gott, J. Richard, A. L. Melott, and M. Dickinson. "The sponge-like topology of large-scale structure in the universe." *Astrophysical Journal*, 306: 341, 1986.

Gott, J. Richard, J. Miller, T. X. Thuan, et al. "The topology of large-scale structure. III—analysis of observations." *Astrophysical Journal*, 340: 625, 1989.

Gott, J. Richard, and M. J. Rees. "A theory of galaxy formation and clustering." *Astronomy & Astrophysics*, 45: 365, 1975.

Gott, J. Richard, and Zachary Slepian. "Dark energy as double N-flation—observational predictions." *Monthly Notices of the Royal Astronomical Society*, 416: 907, 2011.

Gott, J. Richard, and T. S. Statler. "Constraints on the formation of bubble universes." *Physics Letters B*, 136B: 157, 1984.

Gott, J. Richard, and Edwin L. Turner. "Groups of galaxies. III. Mass-to-light ratios and crossing times." *Astrophysical Journal*, 213: 309, 1977.

———. "Groups of galaxies. IV. The multiplicity function." *Astrophysical Journal*, 216: 357, 1977.

Gott, J. Richard, Edwin L. Turner, and S. J. Aarseth. "*N*-body simulations of galaxy clustering. III—The covariance function." *Astrophysical Journal*, 234: 13, 1979.

Gott, J. Richard, and R. J. Vanderbei. *Sizing up the Universe*. Washington, DC: National Geographic, 2011.

Gott, J. Richard, Michael S. Vogeley, Silviu Podariu, and Bharat Ratra. "Median statistics, H_0, and the accelerating universe." *Astrophysical Journal*, 549: 1, 2001.

Gott, J. Richard, David H. Weinberg, and Adrian L. Melott. "A quantitative approach to the topology of large-scale structure." *Astrophysical Journal*, 319: 1, 1987.

Groth, E. J., and P.J.E. Peebles. "*N*-body studies of the clustering of galaxies." *Bulletin of the American Astronomical Society*, 7: 425, 1975.

Gunn, James E., and J. Richard Gott. "On the infall of matter into clusters of galaxies and some effects on their evolution." *Astrophysical Journal*, 176: 1, 1972.

Guth, Alan H. "Inflationary universe: A possible solution to the horizon and flatness problems." *Physical Review D*, 23: 347, 1981.

Hamilton, A.J.S., J. Richard Gott, and David Weinberg. "The topology of the large-scale structure of the universe." *Astrophysical Journal*, 309: 1, 1986.

Hand, N., et al. "Evidence of galaxy cluster motions with the kinematic Sunyaev-Zel'dovich effect." *Physical Review Letters*, 109: 1101, 2012.

Hartle, J. B., and S. W. Hawking. "Wave function of the universe." *Physical Review D*, 28: 2960, 1983.

Hikage, Chiaki, J. Schmalzing, et al. Minkowski functionals of SDSS galaxies I: Analysis of excursion sets. *Publications of the Astronomical Society of Japan*, 55: 911, 2003.

Hikage, Chiaki, Y. Suto, et al. "Three-dimensional genus statistics of galaxies in the SDSS early data release." *Publications of the Astronomical Society of Japan*, 54: 707, 2002.

Hikage, Chiaki, E. Komatsu, and T. Matsubara. "Primordial non-Gaussianity and analytical formula for Minkowski functionals of the cosmic microwave background and large-scale structure." *Astrophysical Journal*, 653: 11, 2006.

Hoyle, Fiona, et al. "Two-dimensional topology of the Sloan Digital Sky Survey." *Astrophysical Journal*, 580: 663, 2002.

Hubble, Edwin. "Cepheids in spiral nebulae." *Popular Astronomy*, 33: 252, 1925.

———. "Extragalactic nebulae." *Astrophysical Journal*, 64: 321, 1926.

———. "A relation between distance and radial velocity among extra-galactic nebulae." *Proceedings of the National Academy of Sciences*, 15: 168, 1929.

———. *Realm of the Nebulae*. New Haven: Yale University Press, 1936.

Hubble, Edwin, and Milton L. Humason. "The velocity–distance relation among extragalactic nebulae." *Astrophysical Journal*, 74: 43, 1931.

Kallosh, R., and A. Linde. "Superconformal generalizations of the Starobinsky model." *Journal of Cosmology and Astroparticle Physics*, 2013(06): 028, arXiv:1306.3214, 2013.

Kallosh, R., A. Linde, and D. Roest. "Superconformal inflationary α-attractors." *Journal of High Energy Physics*, 2013(11): 198, arXiv:1311.0472, 2013.

Kazanas, D. "Dynamics of the universe and spontaneous symmetry breaking." *Astrophysical Journal*, 41: 59, 1980.

Kim, Juhan, Changbom Park, J. Richard Gott, and John Dubinsky. "The Horizon Run *N*-body simulation: Baryon acoustic oscillations and topology of large-scale structure of the universe." *Astrophysical Journal*, 701: 1547, 2009.

Kim, Juhan, Changbom Park, G. Rossi, et al. "The new Horizon Run cosmological *N*-body simulations." *Journal of the Korean Astronomical Society*, 44: 217, 2011.

Kirshner, R. P. "Six groups of galaxies." *Astrophysical Journal*, 212: 319, 1977.

———. *Extravagant universe*. Princeton: Princeton University Press, 2002.

Kirshner, R. P., et al. "A million cubic megaparsec void in Boötes?" *Astrophysical Journal*, 248: L57, 1981.

Komatsu, E., D. N. Spergel, and B. D. Wandelt. "Measuring primordial non-Gaussianity in the cosmic microwave background," *Astrophysical Journal*, 634: 14, 2005.

Krauss, L. M., and M. S. Turner. "The cosmological constant is back." *General Relativity and Gravitation*, 27: 113, 1995.

Kundić, Tomislav, et al. "A robust determination of the time delay in 0957+561A,B and a measurement of the global value of Hubble's constant." *Astrophysical Journal*, 482: 75, 1997.

Linde, A. D. "A new inflationary universe scenario: A possible solution of the horizon, flatness, homogeneity, isotropy and primordial monopole problems." *Physics Letters B*, 108: 389, 1982.

———. "Chaotic inflation." *Physics Letters B*, 129: 177, 1983.

———. "Sinks in the landscape, Boltzmann brains, and the cosmological constant problem." *Journal of Cosmology and Astroparticle Physics*, 0701: 022, 2007.

Markevitch, M., et al. "Direct constraints on the dark matter self-interaction cross-section from the merging galaxy cluster 1E 0657-56." *Astrophysical Journal*, 606: 819, 2004.

Massey, Richard, et al. "Dark matter maps reveal cosmic scaffolding." *Nature*, 445: 286, 2007.

Matsubara, Takahiko. "Analytic expression of the genus in a weakly non-Gaussian field induced by gravity." *Astrophysical Journal*, 434: L43, 1994.

Mecke, K. R., T. Buchert, and H. Wagner. "Robust morphological measures for large-scale structure in the universe." *Astronomy and Astrophysics*, 288: 697, 1994.

Melott, Adrian L., A. P. Cohen, et al. "Topology of large-scale structure. IV. Topology in two dimensions," *Astrophysical Journal*, 345: 618, 1989.

Melott, Adrian, J. Einasto, et al. "Cluster analysis of the nonlinear evolution of large-scale structure in an axion/gravitino/photino-dominated universe." *Physical Review Letters*, 51: 935, 1983.

Moore, Ben, et al. "The topology of the QDOT IRAS redshift survey." *Monthly Notices of the Royal Astronomical Society*, 256: 477, 1992.

Nussbaumer, Harry. "Einstein's conversion from his static to an expanding universe." *European Physics Journal--History*, 39: 37, 2014.

Ostriker, J. P., and L. L. Cowie. "Galaxy formation in an intergalactic medium dominated by explosions." *Astrophysical Journal*, 243: L127, 1981.

Ostriker, J. P., and S. Mitton. *Heart of Darkness*. Princeton: Princeton University Press, 2013.

Ostriker, J. P., and P.J.E. Peebles. "A numerical study of the stability of flattened galaxies: or, can cold galaxies survive?" *Astrophysical Journal*, 186: 467, 1973.

Ostriker, J. P., P.J.E. Peebles, and A. Yahil. "The size and mass of galaxies, and the mass of the universe." *Astrophysical Journal*, 193: L1, 1974.

Ostriker, J. P., and Paul J. Steinhardt. "Cosmic concordance." arXiv:astro-ph/9505066, 1995.

———. "The observational case for a low-density universe with a non-zero cosmological constant." *Nature*, 377: 600, 1995.

Paczyński, Bohdan. "Gravitational microlensing by the galactic halo." *Astrophysical Journal*, 304: 1, 1986.

Paczyński, Bohdan. "Giant luminous arcs discovered in two clusters of galaxies." *Nature*, 325: 572, 1987.

Parihar, Prachi, et al. "A topological analysis of large-scale structure, studied using the CMASS sample of SDSS-III." *Astrophysical Journal*, 796: 86, 2014.

Park, Changbom. "Large *N*-body simulations of a universe dominated by cold dark matter." *Monthly Notices of the Royal Astronomical Society*, 242: P59, 1990.

Park, Changbom, Yun-Young Choi, Sungsoo S. Kim, et al. "The challenge of the largest structures in the universe to cosmology." *Astrophysical Journal*, 759: L7, 2012.

Park, Changbom, Yun-Young Choi, Michael S. Vogeley, et al. "Topology analysis of the Sloan Digital Sky Survey. I. Scale and luminosity dependence." *Astrophysical Journal*, 633: 11, 2005.

Park, Changbom, and J. Richard Gott. "Simulation of deep one- and two-dimensional redshift surveys." *Monthly Notices of the Royal Astronomical Society*, 249: 288, 1991.

Park, Changbom, and Young-Rae Kim. "Large-scale structure of the universe as a cosmic standard ruler." *Astrophysical Journal*, 715: L185, 2010.

Peacock, J. A. "Slipher, galaxies, and cosmological velocity fields." In M. J. Way and D. Hunter, eds., "Origins of the expanding universe: 1912–1932," *Astronomical Society of the Pacific Conference Series*, Vol. 471, 2013.

Peacock, J. A., and S. Cole. "A measurement of the cosmological mass density from clustering in the 2dF Galaxy Redshift Survey," *Nature*, 410: 169, 2001.

Peebles, P.J.E. "Origin of the angular momentum of galaxies." *Astrophysical Journal*, 155: 393, 1969.

———. "The gravitational-instability picture and the nature of the distribution of galaxies." *Astrophysical Journal*, 189: 51, 1974a.

———. "The nature of the distribution of galaxies." *Astronomy and Astrophysics*, 32: 197, 1974b.

———. "Large-scale background temperature and mass fluctuations due to scale-invariant primeval perturbations." *Astrophysical Journal*, 263: L1, 1982.

———. "Dark matter and the origin of galaxies and globular star clusters." *Astrophysical Journal*, 277: 470, 1984a.

———. "The origin of galaxies and clusters of galaxies." *Science*, 224: 1385, 1984b.

———. "Tests of cosmological models constrained by inflation." *Astrophysical Journal*, 284: 439, 1984c.

———. *The Large-scale Structure of the Universe*. Princeton: Princeton University Press, 1993a.

———. *Principles of Physical Cosmology*. Princeton: Princeton University Press, 1993b.

Peebles, P.J.E., and R. H. Dicke. "Origin of the globular star clusters." *Astrophysical Sciences*, 154: 891, 1968.

Percival, Will J., et al. "The 2dF Galaxy Redshift Survey: The power spectrum and the matter content of the Universe." *Monthly Notices Royal Astronomical Society*, 327: 1297P, 2001.

Peebles, P.J.E., L. A. Page, and R. B. Partridge. *Finding the Big Bang*. New York: Cambridge University Press, 2009.

Peebles, P.J.E., and Bharat Ratra. "Cosmology with a time-variable cosmological 'constant.'" *Astrophysical Journal*, 325: L17, 1988.

Perlmutter, S., et al. "Measurements of Ω and Λ from 42 high-redshift supernovae." *Astrophysical Journal*, 517: 565, 1999.

Planck Collaboration: P.A.R. Ade, et al. "Planck 2013 results: I. Overview of products and scientific results." *Astronomy & Astrophysics*, 571: A1, 2014a. (arXiv:1303.5062v2 [astro-ph.CO], 2014)

———. "Planck 2013 results: XVI. Cosmological parameters." *Astronomy and Astrophysics, 571*: 16, 2014b.

———. "Planck 2015: XX. Constraints on inflation." arXiv:15020114, 2015a.

———. "Planck 2015: XIV. Dark energy and modified gravity." arXiv:1502.01590, 2015b.

Pooley, D., et al. "X-ray and optical flux anomalies in quadruply lensed quasars. II. Mapping the dark matter content in elliptical galaxies." *Astrophysical Journal*, 744: 111, 2011.

Rees, Martin J. *Just Six Numbers*. London: Weidenfeld & Nicolson, 1999.

Rees, Martin J., and J. P. Ostriker. "Cooling, dynamics and fragmentation of massive gas clouds: clues to the masses and radii of galaxies and clusters." *Monthly Notices of the Royal Astronomical Society*, 179: 541, 1977.

Riess, Adam G., et al. "Observational evidence from supernovae for an accelerating universe and a cosmological constant." *Astronomical Journal*, 116: 1009, 1997.

Roberts, M. S. "The rotation curves of galaxies." In A. Hayli, Ed., *Dynamics of Stellar Systems* (IAU Symposium 69), pp. 331-340. Dordrecht: Reidel, 1975.

Ryden, Barbara S., et al. "The area of isodensity contours in cosmological models and galaxy surveys." *Astrophysical Journal*, 340: 647, 1989.

Sachs, R. K., and A. M. Wolfe. "Perturbations of a cosmological model and angular variations of the microwave background." *Astrophysical Journal*, 147: 73, 1967.

Sánchez, Ariel G., et al. "The clustering of galaxies in the SDSS-III Baryon Oscillation Spectroscopic Survey: Cosmological implications of the large-scale two-point correlation function." *Monthly Notices of the Royal Astronomical Society*, 425: 415, 2012.

Seldner, M., B. Siebers, E. J. Groth, and P.J.E. Peebles. "New reduction of the Lick catalog of galaxies." *Astronomical Journal*, 82: 249, 1977.

Seo, Hee-Jong, et al. "Acoustic scale from the angular power spectra of SDSS-III DR8 photometric luminous galaxies." *Astrophysical Journal*, 761: 13, 2012.

Slepian, Zachary, J. Richard Gott, and Joel Zinn. "A one-parameter formula for testing slow-roll dark energy: Observational prospects." arXiv:1301.4611 [astro-ph.CO], 2013.

Slipher, V. M. "Nebulae." *Proceedings of the American Philosophical Society*, 56: 403, 1917.

Soneira, R. M., and P.J.E. Peebles. "A non-dynamical hierarchical model of galaxy clustering." *Bulletin of the American Astronomical Society*, 8: 354, 1976.

Speare, Robert, J. Richard Gott, Juhan Kim, and Changbom Park. "Horizon Run 3: Topology as a standard ruler." *Astrophysical Journal*, 799:176, 2015.

Spergel, D. N., et al. "First-year Wilkinson Microwave Anisotropy Probe (WMAP) observations: Determination of cosmological parameters." *Astrophysical Journal Supplement Series*, 148: 175, 2003.

Spergel, D. N., et al. "WFIRST-2.4: What every astronomer should know." arXiv:1305.5425, 2013.

Springel, Volker, Carlos S. Frenk, and Simon D. M. White. The large-scale structure of the universe. *Nature*, 440: 1137, 2006.

Springel, Volker, Simon D. M. White, et al. "Simulations of the formation, evolution and clustering of galaxies and quasars." *Nature*, 435: 629, 2005.

Starobinsky, A. A. "A new type of isotropic cosmological models without singularity." *Physics Letters B*, 91: 99, 1980.

Steinhardt, P. J., and N. Turok. "Cosmic evolution in a cyclic universe." *Physical Review D*, 65: 126003, 2002.

Sunyaev, R. A., and Ya. B. Zeldovich. "The observations of relic radiation as a test of the nature of X-ray radiation from the clusters of galaxies." *Comments on Astrophysics and Space Physics*, 4: 173, 1972.

Thuan, Trinh X., J. Richard Gott, and Stephen E. Schneider. "The spatial distribution of dwarf galaxies in the CfA slice of the universe." *Astrophysical Journal*, 315: L93, 1987.

Tully, R. Brent, Helene Courtois, Yehuda Hoffman, and Daniel Pomarède. "The Laniakea supercluster of galaxies." *Nature*, 513: 71, 2014. (Video presentation at https://www.youtube.com/watch?v=rENyyRwxpHo).

Turner, E. L., and J. Richard Gott. "Groups of galaxies. I. A catalog." *Astrophysical Journal Supplement*, 32: 409, 1976.

Vilenkin, A. "Cosmic strings." *Physical Review D*, 24: 2082, 1981.

———. "Creation of universes from nothing." *Physics Letters B*, 117: 25, 1982.

———. *Many Worlds in One*. New York: Hill and Wang, 2006.

Vishniac, E. T., J. P. Ostriker, and E. Bertschinger. "Explosions in the early universe." *Astrophysical Journal*, 291: 399, 1985.

Vogeley, Michael S., Changbom Park, et al. "Topological analysis of the CfA redshift survey." *Astrophysical Journal*, 420: 525, 1994.

Wainwright, C. L., M. C. Johnson, A. Aguirre, and H. V. Peiris. "Simulating the universe(s) II: Phenomenology of cosmic bubble collisions in full general relativity." *Journal of Cosmology and Astroparticle Physics*, Issue 10: article id. 024, 2014.

Walsh, D., R. F. Carswell, and R. J. Weymann. "0957+561 A,B: Twin quasistellar objects or gravitational lens?" *Nature*, 279: 381, 1979.

Weinberg, David H., Scott Burles, et al. "Cosmology with the Lyman-alpha forest." arXiv:astro-ph/9810142, 1998.

Weinberg, David H., J. Richard Gott, and Adrian L. Melott. "The topology of large-scale structure. I—topology and the random-phase hypothesis." *Astrophysical Journal*, 321: 2, 1987.

Weinberg, Steven. "Anthropic bound on the cosmological constant." *Physical Review Letters*, 59: 2607, 1987.

———. "The cosmological constant problem." *Reviews of Modern Physics*, 61: 1, 1989.

Wells, A. F. *Three-dimensional Nets and Polyhedra*. New York: Wiley, 1977.

White, Simon D. M., Carlos S. Frenk, Marc Davis, and George Efstathiou. "Clusters, filaments, and voids in a universe dominated by cold dark matter." *Astrophysical Journal*, 313: 505, 1987.

Zeldovich, Ya. B. "The theory of the large scale structure of the universe." In M. S. Longair and J. Einasto, eds., *The Large Scale Structure of the Universe: Proceedings of the Symposium, Tallinn, Estonian SSR, September 12–16, 1977* (pp. 409–22). Dordrecht: D. Reidel, 1978.

Zeldovich, Ya. B., J. Einasto, and S. F. Shandarin. "Giant voids in the universe." *Nature*, 300: 407, 1982.

Zumino, Bruno. "Supersymmetry and the vacuum." *Nuclear Physics B*, 89: 535, 1975.

Zwicky, F. "Nebulae as gravitational lenses." *Physical Review*, 51: 290, 1937a.

———. "On the masses of nebulae and of clusters of nebulae." *Astrophysical Journal*, 86: 217, 1937b.

———. "Blue compact galaxies." *Astrophysical Journal*, 142: 1293, 1965.

Zwicky, F., and C. T. Kowal. *Catalogue of Galaxies and of Clusters of Galaxies*, Vol. VI. Pasadena: California Institute of Technology, 1968.

Index

NOTE: Page numbers followed by *f* indicate a figure. Those followed by *t* indicate a table.

Aarseth, Sverre, 61–63, 144
Abbott, Edwin A., 230n2
address in space, 56, 151
adiabatic fluctuations, 60, 61*f*, 66–67, 98–100, 164
Adler, R. J., 124, 187
Adrian, Edgar Douglas, Lord, 56
Aguirre, A., 207
Albrecht, Andreas, 91, 95–97, 207
Alcock, Charles, 36
Alpert, Mark, 228n3
Alpher, Ralph, 41–43, 79, 229n1
American school of topology, 105. *See also* meatball topology
Andromeda Galaxy (M31), 2, 6, 224; approach velocity of, 11–12, 50; dark matter of, 33; distance from Earth of, 9–10; formation of, 49–50; Hubble's photographs of, 8–9; rotational velocity of, 32; shape of, 9
anthropic principle, 201–2
approach velocity, 11
Aristotle, 3
Arkani-Hamed, Nima, 207, 232n3
atomic bombs, 64–65
Aubourg, Eric, 221
axions, 40

Baade, Walter, 29
baby universes, 208–9
Bahcall, Neta, 55–56
Baker, Alan, 56
Banchoff, Tom, 112, 230n2
Bardeen, James, 90, 124, 146
Barrow, John, 201
baryon acoustic oscillations (BAO), 191, 197–98, 216–23, Color Plates 12, 14, 15
Baryon Oscillation Spectroscopic Survey (BOSS), 220–23, 226, Color Plate 12
Bell, Jocelyn, 29
Berlind, Andreas, 167
Bertschinger, Ed, 75–77, 140–41, 151, 231n1(Ch8)
Bhavsar, Suketu, 155–56
BICEP2/Keck and Planck Collaborations, 192
Big Bang model, 3, 22–23, 26, 224; bubble universes and, 94–98, 213–14, 225; cosmic microwave background in, 79–80, 84–85, 89, 180–92, Color Plates 8, 13; early inflationary epoch of, 86–89, 93–94, 98, 145, 210–11; hot model of, 41–43; initial density fluctuations in, 180–84; lookback-time distance in, 228n1; matter-dominant epoch of, 91, 101–2; peculiar velocities in, 150–52, 154; radiation density in, 58–59; radiation-dominant epoch of, 91, 93, 98–99; recombination epoch of, 42–43, 59; transition epoch of, 93–96. *See also* expansion of the universe; inflationary model
Big BOSS, 226
Big Crunch, 19, 23, 82–83, 214, 232n3
Big Rip singularity, 210*t*, 216, 225

Big Science, 155–56
binary galaxies, 52, 59
Binney, James, 49
Blackburne, Jeffrey, 36–37
black holes, 203–4
black hole solution, 14
blueshifts, 11
blue stars, 47–48
Blumenthal, George, 100
Boddy, Kimberley, 206
body-centered cubic array, 106–7
Boltzmann brains, 204–7, 210*t*, 213
Bolyai, Janos, 112
Bond, J. R., 124
Bond, Richard, 164–67
Boomerang, 197–98
Boötes void, 71–72, 139
Born, Max, 106
brightness, 4–5, 8
Broadhurst, T. J., 152–54
Brout, R., 96
bubble topology, 130*f*, 139, 149, 159, 161
bubble universes, 94–98, 213–14, 225
Bullet Cluster, 37–38, 176, Color Plate 3
Burstein, David, 149–50
Burt, Michael, 115
Bush, George H. W., 230n1

Caldwell, Robert, 215–16
Campbell, W. W., 14
Canavezes, A., 162
Carroll, Sean, 206
Carswell, R. F., 34
Carter, Brandon, 201
causal contact, 80–91; causal horizons and, 81*f*, 82–84, 91; in early inflationary epochs, 87–91
CCD digital cameras, 48–49, 157–58
Centaurus, 150
Cepheid variables, 9, 221
CfA Great Wall, 172, Color Plates 4, 8; discovery of, 135–43, 170; survey of, 162–64, Color Plate 6
chaotic inflation, 99, 192
checkerboards, xi–xii
Choi, Yun-Young, 177–79, Color Plate 12
Clinton, Bill, 230n1
cold dark matter (CDM) model, 100–102, 125–28, 129*f*, 179; galaxy formation and, 146; Great Walls in, 142–43, 147–54; in Park's simulation, 145–54
Coleman, Sidney, 93–94
collapse time, 43–44, 45*f*
Colley, Wes, 167, 184, 231n2
Coma Cluster of galaxies, 55, 172, Color Plates 4, 8; elliptical and SO galaxies in, 48; fan-shaped slice map containing, 138–39; infall model for, 46–48, 168; in IRAS Sample, 161*f*; Park's simulation and, 147–48; Supercluster neighborhood of, 152; Zwicky's calculation of mass of, 30–31, 136
Coma Supercluster, 152
compact galaxies, 29
computer simulations. *See* 1D genus simulations; 2D genus simulations; 3D genus simulations
CONTOUR3D, 121–23, 133*f*
Copernican principle, 205, 232n1
Copernicus, Nicolaus, 2, 97
cosmic address, 56, 151
Cosmic Microwave Background Explorer (COBE), 183, 189, 231n2
cosmic microwave background radiation, 43, 79–80, 84–85, 89, 172, 180–92; COBE views of, 183, 189, 231n2; initial density fluctuations in, 180–84; pattern of spots in, 197–98; power spectrum of fluctuations in, 191–92, Color Plate 14; 2D genus topology of, 184–89; WMAP map of, 181–84, 188–89, 231n2, Color Plates 8, 13
cosmic strings, 141–42
cosmic web, xiii, 210*t*, Color Plate 16; dark energy in, 193–226; of the future, 223–26, 233n4; genus statistics and, 222–23; woven filaments of, 164–67, 179, Color Plate 16. *See also* sponge topology; 3D genus simulations
cosmological constant, 14–15, 18, 25, 27, 194–96, 228nn2–3
Courtois, Helene, 151–52
covariance function, 52–54, 59–60
Cowie, Len, 75–77
Coxeter, H.S.M., 112–15
Crab Nebula (M1), 5–6
Curtis, Heber, 5–7, 9
curvature, 121–23

dark energy, 167, 193–226; baby universes and, 208–9; baryon acoustic oscillation measurement of, 216–23, Color Plates 12, 14, 15; Boltzmann brains and, 204–7, 210*t*, 213; constant energy density of, 194–95;

expanding negative-vacuum-energy bubbles in, 206–9, 210*t*, 225, 232–33nn3–4; Gibbons and Hawking's (seen) radiation and, 203–6, 213, 224; increase in the expanding universe of, 196–98, 210*t*, 224–26; local conservation of energy in, 195–96; measuring the value of *w* and, 211–15, 217–22, 225–26, 233n4; new inflationary epoch and, 203; phantom energy and, 209, 210*t*, 215–16, 233n4; recycling of universes and, 209; slow-roll model of, 209–16, 221–22, 233n4; standard form of, 198–202, 210*t*

Dark Energy Spectrographic Instrument (DESI), 226

dark matter, 32–40, 100, 194–97; composition of, 37–40; decrease in density with time of, 195–97; gravitational lensing of, 33–37. *See also* cold dark matter model

Dark Matter and the Cosmic Web Story (Einasto), 134

"Dark matter maps reveal cosmic scaffolding" (Massey et al.), 176

Davies, Roger, 149–50

Davis, Marc, 136, 139–42

Dekel, Avishai, 151, 231n1(Ch8)

de Lapparent, Valerie, 137–40, 147–50, 156

Democritus, 6

density fluctuations, 44–46, 51–63, 81*f*; adiabatic forms of, 60, 61*f*, 66–67, 98–100, 164; after the Big Bang, 180–84; of cold dark matter, 100–102, 125–28, 129*f*; inflation and, 89–94, 103–5, 116–17

de Sitter, Willem, 15–17, 25

de Sitter spacetime, 16–17, 19–21, 26–27, 86, 196

deuterium, 39

de Vaucouleurs, Gérard, 55–56, 71

"The Development of Irregularities in a Single Bubble Inflationary Universe" (Hawking), 97, 100

Dickinson, Mark, xiii, 119–21, 134, 140, 166–67, 230n3

Doppler shifts, 11–12, 182

Doroshkevich, A. G., 60, 61*f*, 72, 124

Dressler, Alan, 149–50

Dumbbell Nebula (M 27), 6

dwarf galaxies, 31

Dyson, Freeman, 100, 213

early-type galaxies, 30–31

Eastwood, J. W., 146

Eddington, Arthur, 13–14, 25

Efstathiou, George, 102, 139–40, 146, 159–61

Einasto, Jaan, 32–33, 134; on distribution of galaxies in space, 71–72; on Voronoi honeycombs, 67–78

Einstein, Albert, 12–20; on the age of the universe, 25; on the atomic bomb, 64; black hole solution of, 14; cosmic string equations of, 142; cosmological constant of, 14–15, 18, 25, 27, 86, 194–96, 228nn2–3; on deflections of light, 13–14; energy-mass equivalence formula of, 58, 60, 65, 83, 229n1; on the expanding universe, 22*f*, 24–25; general relativity theory of, 12–14, 16, 227n1(Ch1), 228n3; on gravity's bending of light, 33–34, 35*f*, 90, 192; on the photoelectric effect, 16; planetary trajectory calculations of, 13; special relativity theory of, 80–81, 86, 195, 202–3; static 3-sphere universe of, 3, 14–20, 24*f*, 25, 86

Einsteinian deflection, 13–14

Eisenstein, Daniel, 217–22

elementary particles, 39–40; exotic forms of, 42–43, 59; Higgs boson, 39, 155–56, 232n3

Elias-Miró, Joan, 207

elliptical galaxies, 30–31, 149–50

Ellis, Richard, 159–61

empty universe, 15–17

energy content of the universe, 195–96

energy-mass equivalence formula, 58, 60, 65, 83, 229n1

Englert, François, 96

equilibrium clusters, 44

EUCLID satellite, 226

Euler, Leonard, 124, 231n4

Euler characteristic, 124–25, 231n4

event horizons, 203–5, 213

exotic elementary particles, 42–43, 59

expansion of the universe, xii–xiv, 2–3; acceleration in, 27, 70, 193–98, 201–4, 232n2; causal connections in, 80–91; dark energy and, 195–226; dark matter and, 33, 195–97; Einstein's views on, 22*f*, 24–25, 27; in the future, 205–6, 224–26, 233n4; Hubble's calculations of, 21–27; inflationary model of, 87–102; linear velocity-distance relations in, 18–27; receding quasar in, 29; in Slipher's table of redshifts, 10–12, 16, 18, 27; spacetime diagram of, 80–83; thermal radiation in, 25–26, 42–43; WMAP chart of, 216–17. *See also* formation of galaxy clusters

The Extravagant Universe (Kirshner), 193–94

Faber, Sandra, 149–50
Farhi, E., 208
Feynman, Richard, 28, 54, 137
filaments, 164–67, 179, Color Plate 16
first cause, 3, 14
Fisher, Richard, 156
five-pointed stars, xii
Flatland: The Movie, 230n2
formation of galaxy clusters, 41–63; adiabatic fluctuations in, 60, 61*f*, 66–67, 164; angular momentum (rotation) and, 49; collapse time in, 43–44, 45*f*; equilibrium in, 44; fractional density enhancement and, 43–44; hierarchical model and neighborliness of, 52–56, 62–63, 73–74, 77, 229n2(Ch3); *vs.* individual galaxies, 49; infall model for, 46–48; intergalactic gas, 46–47, 48; isothermal fluctuations in, 57–63, 66; mass density fluctuations in, 44–46, 51–52; thermonuclear explosion models of, 75–77; Voronoi honeycomb model of, 67–78; X-ray emissions in, 47, 48; Zeldovich pancake model of, 67, 68–69, 71–73, 76–78, 102
Fourier technique, 146, 230n3
fractional density enhancement, 43–44
Freeman, Ken, 32
Frenk, Carlos S., 139–40, 159–61
Friedmann, Alexander, 18–19, 21, 23–27
Friedmann Big Bang universe, 18–19, 21, 23–27, 82–83
fusion reactions, 65
future of the universe, xiii–xiv, 205–6, 224–26, 233n4. *See also* dark energy

galaxies: fluctuations of mass density leading to, 51–52; formation of, 50–52, 60–63, 146; measuring size and distance of, 149–52, 231n1(Ch8); redshifts of, 135
galaxy clusters, xii–xiv, Color Plate 2; cold dark matter fluctuations and, 100–102, 125–28, 129*f*; dark matter in, 32–40, 194–97; formation of, 41–63; gravitational lensing and, 34–37; Great Attractors and, 150–52, 154, Color Plate 5; mass-to-light ratios in, 135–36; neighborliness among, 52–53; in Park's simulation, 146–54; redshifts within, 135–36; superclusters and, 50, 55–56; in thin-slice surveys, 137–40; Zwicky's work on mass of, 30–32. *See also* formation of galaxy clusters
Galileo Galilei, xi, 2, 6, 8
Gammie, Charles, 155–56
Gamow, George, 25–27, 229n1; hot Big Bang model of, 41–42, 79; on radiation density, 58–59, 85
gas clouds, 6
Gauss, Carl Friedrich, 108–9
Gauss-Bonnet theorem, 121–22
Gaussian random-phase distributions, 98–99, 104–5
Geller, Margaret: 3D volume sample of, 162–64; Great Wall discovery by, 136–43, 147–50, 156–57, 172, Color Plates 4, 8
general relativity, 3, 12–13, 79, 227n1(Ch1); on bending of light, 33–34; cosmological constant in, 14–15, 18, 25, 27, 86, 194–96, 228nn2–3; on gravitational attraction, 82–83, 86; on gravitational redshifts, 16; on gravitational repulsion, 195
genus (definition), 119–21, 188*f*
geodesic trajectories, 13
geometry. *See* polyhedrons
The Geometry of Random Fields (Adler), 124
Gibbons, Gary, 97, 203–6, 213, 224
Giovanelli, Riccardo, 155–56, 158–59
Gleick, James, 134
globular clusters, 4–6, 18, 20, 50–52
Goldreich, Peter, 76
googolplex, 203–4
Gott, J. Richard, 43; on bubble universes, 94–98, 213–14; COBE 2D topology project, 189; cosmic string work of, 142; on Einstein's field equations, 228n3; galaxy cluster simulation of, 144, 161; on galaxy cluster sizes, 101; on gravitational lensing, 34–36; infall model of, 46–48; on mass-to-light ratios, 69–70; on median statistics, 70; on pseudopolyhedrons, 112–17; Slepian-Gott-Zinn formula for *w*, 211–15, 217–22; Sloan 3D topology project, 175–79; Sloan Digital Sky Survey, 167–72. *See also* sponge topology
Gott-Melott-Dickinson computer simulation and observations, 116–23, 134, 140, 166–67, 230n3
Gratton, S., 207
gravitational lensing, 34–37, 70
gravitinos, 39, 100
gravity: cold dark matter model and, 101–2, 149; curvature of spacetime and, 82–83; density fluctuations and, 103–5; Einstein's equations of, 33–34, 35*f*, 83, 90, 192, 229n1;

Great Attractor watersheds of, 149–52, Color Plate 5; Newton's theory of, 14, 83, 146, 229n1; of pressure, 83, 86, 229n1
Great Attractors, 149–52, 154, Color Plate 5
Great Repulsors, 150–51
Great Walls: CfA survey of the, 162–64, Color Plate 6; Geller and Huchra's discovery of, 135–43, 170, 172, Color Plates 4, 8; in IRAS Sample, 161*f*, 162; in Park's simulations, 147–49, 152–54; Sloan Great Wall, xiii, 167–72, 174–75, Color Plates 7, 8
Green, Melinda, 115
Guinness Book of Records, 170–71
Gunn, Jim, 49, 65, 67; CCD digital camera design of, 48–49, 157–58; on the Coma cluster, 139; honors and awards of, 48, 63; infall model of, 41–51; on the Milky Way and Andromeda, 49–50; Sloan Digital Sky Survey of, 167–72
Gunn-Peterson effect, 41
Gunzig, E., 96
Guth, Alan, 85–87, 91–97, 197, 208
Guven, J., 208

habitability, 201–4, 210, 213–14
Hahn, Otto, 64
Hale, George Ellery, 7–8
Hambrick, Clay, 175–79
Hamilton, Andrew, 123–24, 155–56
Hand, Nick, 183
Harrison, Edward Robert, 85, 97
Harrison-Zeldovich scale-invariant spectrum, 85, 90, 91, 93, 97–98, 191
Hartle, J. B., 97
Hawking, Stephen, 90, 97, 100, 203–6, 213, 224
Haynes, Martha, 155–56, 158–59
Heggie, Douglas, 56
Heisenberg, Werner, 98
Heisenberg uncertainty principle, 89–90, 98, 191–92
Herman, Robert, 41–43, 79, 229n1
Herschel, William, 6
Higgs boson, 39, 155–56, 232n3
Higgs field, 91, 96–97, 99, 198
Higgs vacuum, 207, 213, 224–25, 232n3
Hikage, Chiaki, 124
Hiscock, William, 142
Hockney, R. W., 146
Hoffman, Yehuda, 151–52
Hofstetter, Lorne, 170
honeycomb topology, 67–78, 128–33, 230–31n3; filaments in, 166–67; three-dimensional simulations of, 74–75; two-dimensional simulations of, 72–74
Horizon Run simulations, 173–75, 223, Color Plate 11
hot model of the Big Bang, 41–43, 79
"How Filaments of Galaxies Are Woven into the Cosmic Web" (Bond, Kofman, and Pogosyan), 164–67
Hoyle, Fiona, 169
Hubble, Edwin, xii, 1–3; Curtis-Shapley controversy and, 7–9; galaxy discoveries by, 10; honors and awards of, 3; linear velocity-distance relation of, 20–27, 35, 43, 70; Mount Wilson telescope and, 8–10; redshift observations by, 11
Hubble constant, 20–27, 35, 43, 70
Hubble Space Telescope, 27, 49, 226
Huchra, John: 3D volume sample of, 162–64; Great Wall discovery by, 136–43, 147–50, 156–57, 172, Color Plates 4, 8
Humason, Milton, 21–22
hydrogen bombs, 64–65

infall model, 46–48
Infinite Polyhedrons (Wachmann, Burt, and Kleinman), 115
inflationary model, xii–xiii, 79–102, 191–92, 227n1(Preface); baby universes in, 208–9; bubble universes in, 94–98, 213–14, 225, 232–33nn3–4; causal contact and, 87–91; chaotic inflation model of, 99, 192; cold dark matter model of, 100–102, 125–28, 129*f*, 179; cosmic strings and, 141–42; density fluctuations in, 89–94, 103–5, 116–17, 180–84; early inflationary epoch in, 86–89, 93–94, 98, 145, 210–11; flat universe models in, 196–98; genus predictions and, 123–25; growth by gravitational instability in the linear regime in, 104; late low-grade inflation in, 193–98, 203, 210–11, 215; multiverse of, 99–100; peculiar velocities in, 150–52, 154; power spectrum in, 145; quantum tunneling in, 92–98, 209; random quantum fluctuations in, 89–94; scalar field of, 91; slow-roll dark energy and, 209–15, 221–22; sponge topology and, 116–34, 193; vacuum energy density of, 91–98, 195, 229n2(Ch5). *See also*

inflationary model (*cont.*)
cold dark matter model; sponge topology; 3D genus simulations
Infrared Astronomical Satellite (IRAS), 159–61
Inner Space/Outer Space conference, 96
integrated Sachs-Wolf effect, 182, 198
intelligent life, 201–4, 210, 213–14
Intel Science Talent Search, 107, 230n1
intergalactic gas, 46–47, 48
intrinsic luminosity, 194
inverse-square law, 4–5, 8
irregular galaxies, 31

Jeans, James, 25
Jefferey, Scott, 155–56
Joeveer, Mihkel, 71–72
Johnson, M. C., 207
Josephson, Brian, 56
jupiters, 36
Jurić, Mario, xiii, 168–71

Kaasik, A., 32–33
Kaiser, Nick, 124, 159–61
Kallosh, R., 192
Kaluza-Klein particles, 40
Kamionkowski, Mark, 215–16
Kant, Immanuel, 6, 9
Kazanas, D., 97
Kepler, Johannes, xi–xii, 110, Color Plate 1
Keynes, John Maynard, 57
Kim, Juhan, 173–75, 177, 178*f*, 223
kinematic Sunyaev-Zeldovich effect, 183
King of Infinite Space (Roberts), 115
Kings College, 57
Kirshner, Robert P., 71, 136, 193–94
Kitt Peak Observatory, 157, 226
Kleinman, Menachem, 115
Klypin, Anatoly, 74–75, 77, 165–67
Kofman, Lev, 164–67
Komatsu, E., 189
Krauss, L. M., 197, 223
Kron, Richard, 157
Kundić, Tomislav, 34–35
Kurzweil, Ray, 107–8

Lange, Andrew, 197–98
Langley, Samuel, 21
Laniakea Supercluster, 151–52, 156, Color Plate 5
Large Hadron Collider, 39
Large Synoptic Survey Telescope (LSST), 226
late-time inflation, 193–98, 203, 210–11, 215
late-type galaxies, 31
Latham, David, 136
Lawrence, Andy, 159–61
Leavitt, Henrietta, 9
Leeuwenhoek, Antoni van, 1
Lemaître, Georges, 26–27, 194, 228n2
Li, Li-Xin, 205
Lick Observatory survey, 54–55
light-years, 1
Linde, Andrei, 91, 95–97; on chaotic inflation, 99, 192; on expanding negative-vacuum-energy bubbles, 206–7, 214; on the multiverse, 100, 232n2; on recycling of universes, 209
linear velocity-distance relations, 18–21; de Sitter effect in, 21; Hubble Constant of, 20–27
Littlewood, John, 56
Lobachevsky, Nikolai, 112
Local Group of galaxies, 56, 71, 224; collapse time of, 50; formation of, 49–50
Local Supercluster, 55–56, 71
lookback-time distance, 228n1
loose star clusters, 6
Lowell, Percival, 77
Lowell Observatory, 10–12
low-grade inflation, 193–98, 203, 210–11, 215
low-matter-density universe, 59–60, 69–70
luminosity: Cepheid variables of, 9, 221; fluctuations of mass density and, 51–52; RR Lyrae stars and, 4–5
luminous red galaxies (LRGs), 177–78, 218–20, Color Plate 15
Lurie, Jacob, 230n1
Lynden-Bell, Donald, 149–50

MACHOS, 36
Magellanic Clouds, 31, 36
Maldacena, Juan, 231n2
Manhattan Project, 64–65
"A Map of the Universe" (Gott, Jurić, et al.), 171–72, Color Plate 8
mapping the universe, xiii, 171–72, Color Plate 8
Massey, Richard, 176
mass of galaxy clusters, 30–31
Mather, John, 183
Matsubara, Takahiko, 168, 177

meatball shift, 168
meatball topology, 55, 63, 67, 76, 78, 103, 105
Melott, Adrian, xiii, 102, 146; measuring the cosmic web project of, 155–56; original topology paper of, 116–21, 134, 140, 166–67, 230n3; on 2D genus topology, 184–87
Messier, Charles, 5
Messier objects, 5–6
microlensing, 36–37
Milky Way Galaxy, xii, 1–2, 119, Color Plates 5, 16; Andromeda's approach towards, 11–12, 50; cosmic address of, 56, 151; dark matter of, 33; formation of, 47, 49–50; galactic center of, 5, 9; globular clusters in, 4–6, 18, 20; gravitational microlensing in, 36; IRAS Sample and, 159, 160*f*; new star formation in, 47; peculiar velocity of, 150; shape and size of, 1, 9, 30; Shapley-Curtis debate on, 6–7, 9; solar system's position in, 2, 4
Millennium-Run simulation, 172–73, Color Plates 9, 10
Miller, John, 155–56
Minkowski functionals, 124, 190
Mollweide projections, 181–82
Moore, Ben, 159–61
Moss, Ian, 97
Mount Wilson telescope, 8–10, 29
M-theory, 200, 225, 232n2, 233n4
multiverse, 94–100

neutralinos, 39
neutrinos, 102
neutron stars, 28
Newton, Isaac, 8, 56–57; gravitational theory of, 14, 83, 146, 229n1; mass formula of, 30; planetary trajectory calculations of, 13; static universe of, 3, 14
Nicholas, Tressilian Charles, 56
nonlinear evolution, 168
Nussbaumer, Harry, 25

octuple galaxies, 52
Oemler, Augustus, 71
1D genus simulations, 189–91
ordinary matter, 38–39, 99–102, 195–98
origins of the universe, xiii. *See also* Big Bang model
Orion Nebula (M42), 6
Ostriker, Jeremiah, 32–33, 134, 144; on the accelerating universe, 196–97; on collapsing galaxies, 49; on matter density, 59–60; thermonuclear explosion model of, 75–77
Ostriker-Bertschinger shell, 75–77

Paczyński, Bohdan, 36
Pagels, Heinz, 100
Palomar Observatory, 3, 27, 48
Pardon, Ruth, 112
Parihar, Prachi, 178–79, 223, Color Plate 12
Park, Changbom, 162–64; COBE 2D topology project and, 189; first simulation by, 144–54; on genus statistics and the cosmic web, 222–23; Horizon Run simulation and, 173–75, Color Plate 11; Sloan Digital Sky Survey and, 167–72
particle physics, 39–40. *See also* elementary particles
Peacock, John, 12, 173
peculiar velocities, 21, 150–52, 154
Peebles, Jim, 32–33, 38, 57; cold dark matter model of, 100–102, 145; on dark energy, 197; on galaxy cluster hierarchies, 52–56, 59–60, 62–63, 73–74, 77–78, 137, 229n2(Ch3); on galaxy rotation, 49; on globular clusters, 50; on mass-to-light ratio, 136; on matter density, 59–60; on slow-roll dark energy, 210
Penzias, Arno, 26, 43
Perlmutter, Saul, 193–94
Perseus-Pisces Supercluster, 152, 158*f*, 159, 161*f*
Petrie, John, 112–15
phantom energy, 209, 210*t*, 215–16, 233n4
Pinwheel Nebula (M101), 6
planar networks, 110–12
planar polygonal tilings, xi–xii
Planck density, 199–202, 216
Planck Satellite Collaboration, 155, 192, 217, 222
planets, xii, 13
Pleiades (M45), 6
Pluto, 12
pocket universes, 99–100
Pogosyan, Dmitry, 164–67
Polk, Kevin, 155–56
Pollack, Jason, 206
polyhedrons, xi–xii, 69, Color Plate 1; regular forms of, 110–12; regular skew forms of, 115; spongelike forms of, xi–xiv, 103–15. *See also* sponge topology
Pomarède, Daniel, 151–52
Pooley, 36–37

Population I stars, 47
Population II stars, 47
power spectrum, 145
power spectrum of fluctuations, 191–92, Color Plate 14
The Prelude (Wordsworth), 57
Prendergast, Kevin, 146
Press, Bill, 50–52, 57, 59, 66, 97
pressure: of radiation, 83, 229n1; of vacuum (dark) energy, 86–87, 194–96, 228n2
Primack, Joel, 100
Principia (Newton), 57
Proxima Centauri, 1, 10
pseudopolyhedrons, 112–17
pulsars, 29

quadruple galaxies, 52
quantum gravity, 214
quantum tunneling, 92–98, 209, 233n4
quasars, 29, 34–37
quintessence models, 210

radiation pressure, 83, 229n1
Rappaport, Saul, 36–37
Ratra, Bharat, 210
recessional velocity, 11–12
recycling of universes, 209
redshifts, 10–12, 16, 18, 27; within clusters, 135–36; within galaxies, 135; Palomar observation of, 27; survey telescope measurements of, 157–58
Rees, Martin, 56–63, 85, 96, 100
reflecting telescopes, 8
regular planar networks, 110–12
regular polyhedrons, 110–12
regular skew polyhedrons, 115
"Rethinking Clumps and Voids in the Universe" (Gleick), 134
Rice, S. O., 189–90
Riess, Adam, 193–94
Roberts, Morton, 32
Roberts, Siobhan, 115
Robertson, Howard P., 24
Roest, D., 192
Rowan-Robertson, Michael, 159–61
RR Lyrae stars, 4–5
Rubin, Vera, 32
Russell, Henry Norris, 18, 34
Ryden, Barbara, 189–91

Saar, Enn, 32–33, 71–72
Sachs-Wolfe effect, 182
saddle-shaped surfaces, 23–24, 107–10
Sagittarius, 5, 9
Sandage, Alan, 136
Saunders, Will, 159–61
Schechter, Paul: on Boötes void, 71–72; Press/Schechter formula of, 36–37, 50–52, 57, 59, 66
Schiaparelli, Giovanni, 77
Schmidt, Brian, 193–94
Schmidt, Maarten, 29
Schneider, Stephen, 155–56
Schwarz P surface, 115
Schwarzschild, Karl, 14
Seaborg, Glenn, 108, 230n1
Seven Samurai, 149–50
Shandarin, Sergei, 67–78, 166–67
Shane, Donald, 54–55
shapes of galaxies, 30–33
Shapley, Harlow, 2–9, 18
Shapley Supercluster, 152
Shectman, Steve, 136
sibling bubble universes, 214, 225, 232–33nn3–4
Silk Mass, 66
Sirius, 1
Slepian, Zachary, 210–12, 216, 221
Slepian-Gott-Zinn formula for *w*, 211–15, 217–22
Slipher, Vesto, 10–12
Sloan Digital Sky Survey, 35, 49, 155, Color Plate 11; CCD digital camera for, 48–49, 157–58; on dark energy, 167, 198; LRGs in, 177–78, 218–20, Color Plate 15; Sloan 3D topology project of, 175–79, Color Plates 11, 12; Sloan Great Wall in, xiii, 167–72, 174–75, Color Plates 7, 8; Sloan III of, 178–79, 219–20, 224, Color Plate 12
slow-roll dark energy, 209–16, 221–22, 233n4
slow-roll inflation, 231n2
Smoot, George, 183
SO galaxies, 30–31, 48
solar system, 2, 4; cosmic address of, 56, 151; individual velocity of, 12
Sombrero Galaxy, 172, Color Plate 8
Soniera, Ray, 52, 55–56, 73–74
South Pole Telescope, 192
Soviet Union: nuclear weapons of, 64–65; school of cosmology of, 57; school of topology of, 105, 134. *See also* Swiss cheese topology

space address, 56, 151
Speare, Rob, 223
special relativity: spacetime diagram of the Big Bang and, 80–81; on vacuum states, 86, 195, 202–3, 229n2(Ch5)
Spergel, D. N., 189
SPIDER high-altitude balloon, 192, 197–98
Spindel, P., 96
spiral galaxies, 30–31, 47. *See also* Milky Way Galaxy
spiral nebulae, 5–6, 11–12, 18
Spitzer, Lyman, 226
spongelike polyhedrons, xi–xiv, 103–15; curvature of, 122–23; pseudopolyhedrons and, 112–17; saddle-shaped surfaces of, 107–10; Schwarz P surface of, 115; semiregular types of, 115
"The Sponge-like Topology of Large-scale Structure in the Universe" (Gott, Melott, and Dickinson), 116–21, 140, 166–67
sponge topology, xiii, 105, 116–34, 193; in cold dark matter simulations, 125–28, 129*f*; dwarf and low-surface-brightness galaxies and, 156; filaments in, 164–67, Color Plate 16; in Geller and Huchra's thin-sliced universe, 138–43, 149; genus predictions for, 123–25, 222–23, 231n4; Gott-Melott-Dickinson computer simulation of, 116–23, 134, 140, 166–67, 230n3; Great Walls in, 139–49, 152–54, 162–64, Color Plates 4, 8; measuring density contour in, 121–23, 133*f*, 158–59; naming of, xiii, 164–67; redefinition of genus in, 119–21; Swiss-cheese simulations and, 128–33; 2D genus view of, 187–89; Virgo Supercluster observational sample of, 119, 128. *See also* cosmic web; 3D genus simulations
Springel, Volker, 144, 172–73, Color Plates 9, 10
standard dark energy, 198–204, 210*t*
Starobinsky, Alexei, 97, 192
starred polyhedrons, xii, Color Plate 1
Statler, Tom, 96, 213–14
Steinhardt, Paul, 91, 95–97; on the accelerating universe, 196–97, 232n2; quintessence model of, 210
string theory, 39–40, 199–202, 212–13
Sun, 2; luminosity of, 4; peculiar velocity of, 21
Sunyaev, R., 60, 61*f*, 70
Sunyaev-Zeldovich effect, 182–83
superclusters, 50, 55–56, 151–52, 156, Color Plate 5
supernovae, 7; explosions of, 29; gas ejecta of, 5–6; intrinsic luminosity of, 194; type Ia of, 194, 217, 220–21, 226; Zwicky's discoveries of, 29
superstring theory, 39–40
supersymmetric partners, 39–40, 212–13
supersymmetry, 39–40
Swiss cheese topology, 67–70, 103, 105, 128–33, 138–43
Szalay, A. S., 124
Szilard, Leo, 64

theory of everything, 39–40, 225–26, 233n4
theory of general relativity. *See* general relativity
theory of inflation. *See* inflationary model
theory of special relativity. *See* special relativity
theory of supersymmetry, 39–40
thermal radiation, 25–26, 42–43; adiabatic fluctuations of, 60, 61*f*, 66–67, 98–100, 164; early fluctuations in, 57–63; energy density of, 58; Gibbons and Hawking (seen) form of, 203–6, 213, 224; gravitational pressure of, 83, 229n1
thermonuclear explosion model, 75–77
thin-slice surveys, 137–40
Thomson, J. J., 14
3D genus studies, 144–79, 181, 222–23, 231n1-(Ch10); CfA Great Wall, 135–43, 170, 172, Color Plates 4, 8; of galaxies, 149–52, 231n1(Ch8); Giovanelli and Haynes sample, 158–59; Great Attractors in, 149–52, 154, Color Plate 5; Great Wall region, 162–64; Horizon Run, 173–75, 223, Color Plate 11; IRAS Sample, 159–61, 159–62; "A Map of the Universe" (Gott and Jurić), 171–72, Color Plate 8; Millennium-Run, 9, 10, 172–73, Color Plates 9, 10; by Park and Gott, 145–54; rodlike samples in, 152–54, 170; Sloan 3D topology project, 175–79, Color Plate 11; Sloan Great Wall discovery, xiii, 167–72, 174–75, Color Plates 7, 8; weblike filaments and, 164–67, 179, Color Plate 16
3-sphere universe, 14–19; de Sitter spacetime and, 16–17, 19–21, 26–27, 86, 196; Friedmann Big Bang universe and, 18–19, 21, 23–27, 82–83; linear velocity-distance relations in, 18–20
Three Dimensional Nets and Polyhedra (Wells), 115
Thuan, Trinh, 155–56

Time Travel in Einstein's Universe (Gott), 232n2
Tipler, Frank, 201
Tolman, Richard C., 20
Tombaugh, Clyde, 12
Tomer, R. H., 146
Tonry, John, 136
topology (definition), 119–21
Trinity College, 56–57
Truman, Harry, 29, 64
Trumpler, R., 14
truncated octahedrons, 106–11
Tully, R. Brent, 151–52, 155–56, 161
Turlevich, R., 149–50
Turner, Ed, 34–35, 59, 61–63; galaxy cluster simulation by, 144, 161; on galaxy cluster sizes, 101; group catalogue of, 135–36; on mass-to-light ratios, 69–70
Turner, M. S., 197
2D genus simulations, 181–89, 231n1(Ch10); Doppler effects in, 182; WMAP map, 181–84, 188–89, 231n2, Color Plates 8, 13
2MASS survey, Color Plate 16
type Ia supernovae, 194, 217, 220–21, 226

uncertainty principle, 89–90, 98, 191–92

velocity dispersion, 149–50
Vilenkin, Alex, 97, 209, 232n1
Virgo Cluster, 55–56, 71; concentration of galaxies in, 150; recessional velocity of, 21
Virgo Supercluster, 55–56, 71, 119; flat structure of, 72–73, 75; Laniakea Supercluster and, 151–52, 156, Color Plate 5
virialization, 45*f*
Vogeley, Michael, 155–56, 162–64, 169; CfA2 Survey and, Color Plate 6; Sloan 3D topology project and, 175–79, Color Plates 11, 12
Voronoi honeycombs, 67–73, 128–33, 139, 230–31n3. *See also* honeycomb topology

Wachmann, Avraham, 115
Wainwright, C. L., 214
Walsh, D., 34
Wambsganss, Joachim, 36–37
Wandelt, B. D., 189
Wavell, Bruce, 105
weakly interacting massive particles (WIMPS), 38–39, 100–101
Wegner, Gary, 149–50
Weinberg, David, 155–56; CONTOUR3D program of, 121–23, 133*f*; IRAS Sample and, 159–61; on 1D genus topology, 190; Sloan Digital Sky Survey and, 167–72, 221
Weinberg, Nevin, 215–16
Weinberg, Steven, 100, 201, 213
Wells, A. F., 115
Westinghouse Science Talent Search, 107–8, 230n1
Weyl, Herman, 19–21
Weymann, R. J., 34
Wheeler, John, 137
Whirlpool Galaxy, 172, Color Plate 8
Whirlpool Nebula (M57), 6
White, Simon, 139–40
Wide Field Infrared Survey Telescope (WFIRST), 226
Wilson, Robert, 26, 43
Wirtanen, Carl, 54–55
Witten, Ed, 200
WMAP satellite team, 155; cosmic microwave background map by, 181–84, 188–89, 231n2, Color Plates 8, 13; expansion of the universe chart by, 216–17; on the value of *w*, 221–22
Wordsworth, William, 57
World War II, 64–65
wormholes, 208–9
Wright, Thomas, 6
Wright brothers, 21

Yahil, Amos, 32–33, 59–60
Yerkes Observatory, 7–8

Zeldovich, Yakov, 57–78; on adiabatic fluctuations, 60, 61*f*, 66–67, 72, 99, 164; Boötes void and, 71–72; on density fluctuations, 57–63, 81*f*; on galaxy clusters at vertices, 69–70; hydrogen bomb work of, 64–65; on median statistics, 70; pancake model of, 67, 68–69, 71–73, 76–78, 102; scale-invariant spectrum of fluctuations of, 85, 90–91, 93, 97–98, 191; Voronoi honeycomb model of, 67–78, 128–33, 139, 166–67
Zinn, Joel, 221–22

Zumino, Bruno, 212
Zwicky, Fritz, xii, 28–33, 40; catalogue of galaxies of, 135; on dark matter, 32–33, 100, 194; on gravitational lensing, 33–37, Color Plate 2; honors and awards of, 29; on mass of the Coma Cluster of galaxies, 30–31, 136; on neutron stars, 28–29; supernovae discoveries by, 29